Collaboration in Libraries and Learning Environments

Collaboration in Libraries and Learning Environments

Editor

Dominika Dechiel

Collaboration in Libraries and Learning Environments

Edited by **Dominika Dechiel**

ISBN: 978-1-68117-253-8
Library of Congress Control Number: 2016934796

© 2017 by
SCITUS Academics LLC,
www.scitusacademics.com
Box No. 4766, 616 Corporate Way,
Suite 2, Valley Cottage,
NY 10989

Notice

Preface

As course management systems have become a popular support to distance learning on campuses, integration of library presence into the courseware "has had a challenging agenda" to academic librarians. This challenge applies especially to those with related position titles, such as a Distance Learning Librarian. It has been recognized that academic librarians "must seek to integrate their resources into online courses delivered via course management systems, in order to ensure that libraries continue to remain vital to higher education." Collaborations are crucial to ensure successful integration of the library into the course management system. The changing environment in higher education requires different approaches to be taken to the provision of professional support services. This may result in the development of outsourced shared services, the convergence of many different student-facing services or the development of more active collaborative networks. This book, Collaboration in Libraries and Learning Environments, reflects the changing context and broad principles affecting the ways in which we need to manage and provide services and offers case studies of changes that have already taken place. This book distinguishes and exposes the innovations that leaders and practitioners are implementing to transform and develop the provision of sustainable and creative support services. Such innovations are resulting in diverse models of service delivery and the development of more active collaborative networks and commercial partnerships.

Table of Contents

Chapter 1   Re-Envisioning Libraries for Training and Literacy
             Development   1

Chapter 2   School Libraries in Greece Turbulent Past, Uncertain
             Present, Doubtful Future   11

Chapter 3   Redefining Library Learning Facilities in Malaysia: Lesson
             from Frank Lloyd Wright Sustainable Approach in Spatial
             and Landscape Design   29

Chapter 4   'In-formation' of Better Learning Environments - the
             Educational Role of the University Library   43

Chapter 5   Data standardization in Digital Libraries: An ETD Case in
             Turkey   73

Chapter 6   Environmental Friendly School Libraries as Excellence
             Resource Center in Creating Human Capital and Learned
             Malaysia Young Generation   87

Chapter 7   Towards a better Design: Physical Interior Environments of
             Public Libraries in Peninsular Malaysia   105

Chapter 8   Social Navigation For Educational Digital Libraries   129

Chapter 9   Qualitative Assessment of Mould Growth for Higher
             Education Library Building in Malaysia   149

Chapter 10   Service Level Agreements for the Digital Library   165

Chapter 11   An Exploratory Study of Collaboration in New Zealand
              Tertiary Libraries   181

Chapter 12   The Adaptable Cycle of Engagement: A Win/Win Model for
              the Library 199

Chapter 13   User-Experience Design and Library Spaces: A Pathway to
              Innovation?   207

Chapter 14   Academic Education in Library and Information
              Management in Bulgaria   227

Chapter 15   Mapping of an Aptitude of Library Science students in
              Relation to Some Variables   251

              Index   261

Chapter 1

Re-Envisioning Libraries for Training and Literacy Development

Kurtis McDonald

Kobe College, 4-1 Okadayama, Nishinomiya, Japan

ABSTRACT

As technology continues to develop at a rapid pace, libraries need to embrace a more engaged role in the active training of patrons to meet the diverse range of literacy requirements they are increasingly likely to face in their daily, academic, and professional lives. This paper contends that the promotion of educational opportunities should run throughout all aspects of libraries and embody the ethos of the librarians that work within them. If libraries are going to truly meet the needs of their patrons going forward, innovative information literacy training must be re-envisioned as a service priority in all library functions.

Keywords: literacy; libraries; computer literacy; technological fluency; information literacy

INTRODUCTION

Libraries have long been considered primarily as places where books, journals, and other forms of paper-based sources of information are stored, indexed, and made available to meet the specific information needs of their patrons as they might arise. While the scope of these collections naturally varies within the different library contexts (public, school, academic, special) and even among the many subdivisions within each of these contexts, the notion that libraries are mainly for making paper-based reading materials accessible is one that has existed as long libraries have existed themselves. In recent decades, however, as technological capabilities have progressed dramatically, so has the capacity of libraries to incorporate an ever-increasing amount of text-based digital formats (electronic journals, magazines, newspapers, e-books) as well an array of multimedia formats (audio, audiobooks, video, video games) in their collections. Although the inclusion of such formats in library collections has served to greatly expand library access to a growing range of information and entertainment sources, it seems that in many ways the focus has managed to remain chiefly centered around that of providing access itself. Though this approach harkens back to the concept of libraries as information repositories carried over from the paper-based age, it is imperative to reconsider if this area of service is all that libraries can and should be concerned with.

In an age when access to information of all kinds is becoming increasingly easy due to the Internet, a library focused on providing access alone is clearly not meeting all of the varied needs of its patrons including a range of literacy needs. Indeed, along with the technological developments that have allowed libraries the opportunity to provide greater access to an expanding range of formats, the current skill set required of people in the 21st century seems to have changed as well. No longer is basic literacy, the ability to read and write, all that is required to thrive and succeed. Today, people must also increasingly be able to use technology effectively; they must have computer literacy. Even beyond basic and computer literacy, people must be able to conduct and manage their own information-seeking processes. Identifying information needs, exploring search options, and evaluating the sources that they do find are all now essential elements of information literacy required of many in the Internet-based, information-rich world of today. While the importance of these literacies has long been recognized within the field of library and information science and many, if not most, libraries do attend to them to some degree, this area of library service remains in a secondary role to that of providing access. If librarianship is going to truly meet the needs of patrons today, innovative information literacy training must be made a service priority in all library functions.

LITERACIES OLD AND NEW

While the term literacy once referred almost exclusively to the basic ability to read and write, the concept of literacy today has come to mean a great deal more. *Basic literacy*, however, remains a critical first step that must be reached before computer and information literacy can be fully addressed. Fortunately, in many parts of the world, the development of functional basic literacy is almost taken for granted, as it is traditionally cultivated in required primary schooling. For those who may not have achieved basic literacy through schooling, both school and public libraries in the U.S. often still provide some degree of support (Brey-Casiano, 2006). In conjunction with other community-based organizations, these and other programs such as English as a Second Language (ESL) instruction remain small but important areas of service provided by libraries.

As technology has become a more and more ubiquitous part of daily life for many people, the ability to use computers effectively has become another critical skill in the 21st century. While the term *computer literacy* is often used to denote some degree of mastery of an overall set of computing skills or abilities in relation to the computer such as being able to use word processing or spreadsheet software, perhaps a more thorough and nuanced understanding is to think of this realm of literacy as *technological fluency* instead (Snyder et al., 1999). Indeed, the word fluency seems to better characterize both the range and fluid nature of technological abilities. While someone may be proficient at using the current version of Microsoft Word, that doesn't necessarily ensure that they would be able to effectively use other types of word processing software or even the next iteration of Word. Most notably promoted in the National Academy of Sciences'

Committee on Information Technology Literacy's 1999 report, the concept of technological fluency not only takes into account the disparities in a computer user's skills between different applications, it also encapsulates the need "to plan to adapt to changes in the technology" (Snyder et al., 1999, p. 2), an aspect that is critically important given the rapid pace of technological change. Overall, the Committee noted that three kinds of knowledge were essential to technological fluency: contemporary skills (being able to use current applications effectively), foundational concepts (having an understanding of how the technology works), and intellectual capabilities (being able to reflect on and apply the use of technology to suit changing needs) (Snyder et al., 1999). While more and more people have now grown up accustomed to using computers in many different capacities, there regrettably remains a substantial 'digital divide' between people who have unfettered access to computers and the Internet in their home, school, and work environments and those that simply can not afford such access. For these people, much like those who may not have achieved basic literacy, public libraries may offer one of the only options available for improving this area of fluency (Celano & Neuman, 2010; Krebeck, 2010) By providing access to public use computers and training sessions geared around specific applications, libraries are now increasingly at the forefront of technological fluency development.

While it is clear that libraries have some role to play in fostering both basic literacy and technological fluency, perhaps their greatest role should be seen in terms of supporting information literacy. *Information literacy*, probably most authoritatively defined by the American Library Association (ALA) in their 1989 report, is generally considered to be the set of abilities required "to recognize when information is needed and the ability to locate, evaluate, and use effectively the needed information" (para. 3). These abilities, therefore, can be thought of as taking people throughout the entirety of the research process: from the initial impetus which sparks a recognition of the need for information, to the search process itself and the evaluation of sources, all the way through to the proper and ethical use of any information found. While obviously of great importance in academia where the ability to conduct independent research and incorporate any relevant background information found is a hallmark of the educational endeavors typically undertaken by students, information literacy extends well beyond coursework and into the daily lives of all of us. Summarizing the widespread importance of information literacy, the ALA (1989) put it this way: "To promote economic independence and quality of existence, there is a lifelong need for being informed and up-to- date" (para. 2).

Although the ALA's 1989 overview of information literacy remains highly relevant and often cited, more recent literature has attempted to recalibrate information literacy to take into account new literacies that are still emerging, especially since many aspects of technological fluency have now become so intertwined with these latest skill sets. Some such as *digital literacy* simply seek to extend the central concepts of information literacy more specifically and directly to the vast number of new information and communication options now available online (Adeyemon, 2009; Borawski, 2009). Another concept gaining notoriety in the literature, *metaliteracy*, expands on the definition of information literacy somewhat by highlighting the growing degree of participatory opportunities now available, as Mackey & Jacobson (2011) note, "to actively produce and share content through social media and online communities" (p. 76). Finally, another conceptualization seen recently, *transliteracy*, focuses on the importance of the interaction now possible between all of the different types of text and multimedia formats, as they are increasingly likely to be used in conjunction with one another rather than in isolation (Ipri, 2010). While such novel frameworks have yet to be fully embraced by the library and information science community as broadly as information literacy has been, the distinctions they raise serve as excellent opportunities to reconsider the further blurring of the line between technological fluency and information literacy.

RETHINKING THE PRIORITIES OF LIBRARIES

While it's clear that the importance given to information literacy in the field of library and information science has grown alongside greater access to information and heightened demand for technological fluency, both in the literature on the subject and in the parallel increase in instruction done in libraries in recent years (Grassian & Kaplowitz, 2009), it's even more clear that this function still remains subordinate to the traditional view of libraries as collections of materials (Jankowska & Marcum, 2010). Though some have called for a dramatic realignment of library priorities to place the central emphasis on instruction, particularly in academic contexts (see Loesch, 2010, and Palmer, 2011), these voices currently remain few and far between. While many have recognized the critical crossroads that libraries find themselves at today and have expressed a range of possible scenarios leading either more towards the margins or the mainstream (see Sennyey, Ross, & Mills, 2009), it seems that the library's potentially preeminent role as a provider of both information and the training required to utilize it effectively is often overlooked.

Given both the noted misalignment of libraries' current priorities retaining their focus foremost on providing access to a traditional collection and the demonstrated need for fostering information literacy skills outlined previously, it is my view that libraries must embrace a greater role for integrated training and instruction in concert with a more innovative approach to the materials they offer. However, beyond offering isolated information instruction sessions, re-envisioned libraries should build instructional opportunities into everything they do and the   materials that they offer. Reference transactions should be thought of as training sessions in their own right and carried out as such as much as possible. Furthermore, the growing potential of digital convergence should be leveraged so that materials themselves can optionally serve as their own interactive training materials combining text with all kinds of multimedia formats. For instance, imagine the potential of downloadable e-books embedded with accompanying audio, video, interactive games, and social networking options. The unique possibilities offered by advanced cloud-based computing systems could also provide entire web-based and on-demand planned learning programs consisting of such multimedia-rich, electronic training materials. These programs could be set up to recognize users' knowledge states and previously viewed content and automatically tailor subsequent materials precisely to the level, interests, and identified needs of the users at that moment. Although the traditional focus of libraries was on ensuring access to carefully selected collections, the re-envisioned focus that I am proposing should be centered on ensuring access to engaging educational materials that foster information literacy and technological fluency in addition to mastery of the content itself.

While it is a liberating intellectual activity to completely re-envision libraries from the ground up, a more measured, incremental progression from a collection focus to a training focus would ultimately likely be required given the realities faced by most libraries. Budget concerns and their impact on limiting how many electronic subscription services could be brought online are obviously one of the largest obstacles to be addressed. An even more important initial concern is gauging the preferences and opinions of any library's user base before any major reprioritization of library goals is enacted. In the end, a library's primary mission must align with providing its constituency with the services that they desire and demonstrate a proclivity toward using. If interactive, electronic multimedia geared toward providing individualized content and training did not match well with the particular library and information needs and preferences of a given library community, such services, no matter how innovative, would likely not be used. Furthermore, as Godwin (2009) notes, the great majority of library faculty and staff are not currently prepared "to employ more engaging and active methods to reach their patrons" (p. 4), as would be required to meet the goals of the re-envisioned library outlined above.

CONCLUSION

Although the future of libraries remains uncertain, it is clear that a significant shift from the overarching priority given to collection of paper-based materials to include a much wider array of electronic formats and Internet resources is already well underway. While most of the field has been content to simply expand the focus on maintaining access to an indexed collection of both paper-based and electronic materials, this paper argues that a much more substantive reconsideration of the primary roles that libraries should be engaged in is required due to both the information and computer literacy needs increasingly required in the Information Age and the new affordances that technology can increasingly offer. Libraries today need to embrace a more engaged role in the active training of patrons to meet the requirements they are more and more likely to face in their daily, academic, and professional lives. The promotion of educational opportunities should run throughout all aspects of libraries and should embody the ethos of the librarians that work within them. While such a drastic realignment is sure to face challenges and resistance from those who do not believe such services can or should fall within the purview of libraries, the realities of the information needs of patrons today demands this type of innovation in strategic planning.

REFERENCES

1. Adeyemon, E. (2009). Integrating digital literacies into outreach services for underserved youth populations. *The Reference Librarian, 50*(1), 85- 98.

2. American Library Association. (1989). Presidential committee on information literacy: Final report. Retrieved from http://www.ala.org/ala/mgrps/divs/acrl/publications/whitepapers/presidential.cfm

3. Borawski, C. (2009). Beyond the book literacy in the digital age. *Children and Libraries, 7*(3), 53-55.

4. Brey-Casiano, C. A. (2006). From literate to information literate communities through advocacy. *Public Library Quarterly, 25*(1/2), 181-190. Celano, D., & Neuman, S. (2010). How to close the digital divide? Fund public libraries. *Education Week, 29*(28), 33.

5. Godwin, P. (2009). Information literacy and Web 2.0: Is it just hype? *Program: Electronic Library and Information Systems, 43*(3), 264-274. Grassian, E. S., & Kaplowitz, J. R. (2009). *Information literacy instruction: Theory and practice (2nd ed.).* New York: Neal-Schuman Publishers,Inc.

6. Ipri, T. (2010). Introducing transliteracy: What does it mean to academic libraries? *College & Research Libraries News, 71*(10), 532-533, 567.

7. Jankowska, M. A., & Marcum, J. W. (2010). Sustainability challenge for academic libraries: Planning for the future. *College & Research Libraries, 71*(2), 160-170.

8. Krebeck, A. (2010). Closing the "digital divide": Building a public computing center. *Computers in Libraries, 30*(8), 12-15. Loesch, M. F. (2010). Librarian as Professor: A dynamic new role model. *Education Libraries, 33*(1), 31-37.

9. Mackey, T. P., & Jacobson, T. E. (2011). Reframing information literacy as a metaliteracy. *College & Research Libraries, 72*(1), 62-78.

10. Palmer, C. (2011). This I believe...All libraries should be teaching libraries. *Portal: Libraries and the Academy, 11*(1), 575-582.

11. Snyder, L., Aho, A. V., Linn, M., Packer, A., Tucker, A., Ullman, J., & Van Dam, A. (1999). *Being Fluent with Information Technology*. The National Academies Press. Retrieved from http://www.nap.edu/openbook.php?isbn=030906399X

12. Sennyey, P., Ross, L., & Mills, C. (2009). Exploring the future of academic libraries: A definitional approach. *The Journal of Academic Librarianship, 35*(3), 252-259.

Chapter 2

School Libraries in Greece Turbulent Past, Uncertain Present, Doubtful Future

Georgios D. Bikos[1][a], Panagiota Papadimitriou[b]

[a,b] Technological Educational Institute of Athens, Department of Library Science and Information Systems, St. Spyridon Str., Egaleo, 12210, Greece

ABSTRACT

School libraries in Greece are officially established as a separate section of schools; a section that has been functioning in cooperation with schools and supporting them on a regular basis, in a well-organised way, employing its own staff since 1995. However, reports and relevant legislation have appeared since the first half of the 19th century. This paper presents and evaluates the legislation that concerns the establishment and operation of school libraries in Greece from the establishment of the Greek state to the present day. Furthermore, reference is made to the extent the relevant legislation was applied in practice and the method employed for its application; at the same time we do not fail to make a brief reference to the first legislative action taken in relation to school libraries in Europe.

Keywords: Libraries; School Libraries; Greece; Legislation; History

INTRODUCTION

"A review of the literature shows that not only is there no comprehensive published history of school libraries[2], but that school libraries are inadequately covered in the general histories of education and librarianship. Those writers who do discuss the history of school libraries tend to assume that they are a more recent phenomenon than they actually are; indeed, some assume they are a 20th-century development. [...] In addition, as Stott pointed out, both general histories of education and histories of libraries and librarianship failed to provide coverage of the historical development of school libraries. [... However,] school libraries have existed in schools since at least the 8th century[3]. These early school libraries would have been very different from school libraries today, just as schools now are very different from their predecessors of earlier centuries." [2] Moreover, it is true that until the early decades of the 20th century, the operation of school libraries was much less underlined by their *school* character than it is today and they were, in effect, not so closely related to the school's educational and instructive character. This is because until then school libraries were often established for other purposes unrelated (or minimally related) to school instruction, as, for instance, meeting the recreational reading needs of children or the needs of the local community [2].

To be more specific about the time when the first school libraries appeared and refer to the legislation governing them, note that, according to historians studying educational legislation, the first decree related to school libraries was issued in England in 1578 and provided that each school should have a library [2, 3]. "Despite all this, there is considerable evidence for the existence of libraries in schools in England (and in some other

European countries) from at least the 8th century; in the United States from the 18th century." [2] When it comes to school libraries in the United States, there is evidence that their development was parallel with that of public libraries and until 1913 the number of working school libraries came to 10.000, but few of them had more than 3.000 volumes/books. However, the first educated school librarian who worked in a school library in the United Stated appears in 1900 [3].

Consequently, we are initially given the impression that Europe established the legal framework for the foundation (and operation) of school libraries much earlier than the United States. The situation does not appear to be uniform throughout all European countries though, e.g. "the history of school libraries in Dalmatia dates back to the 19th century, when they were established as differentiated and defined subjects within the education system in the Austro-Hungarian monarchy. [... More specifically, school] libraries were founded after the adoption of the 1871 Act on Principles of Primary School Education, temporary school and teaching regulation of 1870, and the 1871 Regulations on School Libraries." [4] In Dalmatia, however, they were established "as a tool for the further education of teachers", and not as much as to cover the needs of students. What we should note, however, is that the legal enactment of the operation of school libraries was considered to be part of an effort to modernise the education system [4]. That was the case in almost all countries though.

MAIN PART

Let's take a look at the history of school libraries in Greece and the history of the legislation governing them from the establishment of the Greek state onwards4. Important libraries are found in Greece since the period of the Turkish occupation; these are connected with intellectuals who supported them offering rich collections of books, mainly hand-written. This is how major libraries were established and operated in modern Greece, such as that of Zagora, Dimitsana, Chios, Kozani and elsewhere [10]. Some of them worked also as school libraries within renowned schools that flourished in the 18th and early 19th centuries (e.g. School libraries started operating in secondary education schools in the Peloponnese during the Turkish occupation [11].) These libraries contributed to the dissemination of the spirit of modern Greek Enlightenment – i.e. to the idea that education is directly related to a nation's conquest of liberty, and also to its general prosperity.

Moving on to the years that marked the establishment of the Greek State, we note that the first governor (since 1827), I. Kapodistrias, gave priority to elementary education and vocational training as part of his effort to help the illiterate and unskilled population. He also established the Library of the Central School (an upper level school comprising three years of studies) in Aegina, an island near Attica. This library was considered to be a school library, but also served a more complex role, since it worked as a reading room, a book distribution centre for schools throughout the country and also as a storeroom for school books unsuitable for use in order to be substituted by more useful ones [11]. The Library of the Central School, however, was not the only school library of that period. The other existing libraries, though, were neither in full operation nor did they form a *functional* educational place. A characteristic example of the way school libraries worked at the time is the case of the Cyclades. Out of the four libraries of the Greek schools on Kea, Mykonos, Naxos and Sifnos, the last three were never integrated into the functional school area and therefore remained almost or entirely inactive. As far as the one established on the island of Kea is concerned, its pre-existing core of books compiled in the 17th century was only enriched with the school books distributed throughout Greece [11].

Nevertheless, the institutional establishment of the right development process for school libraries was something that the intellectuals of the time clearly pursued. The attitude adopted by the Governor's direct associates, such as Konstandas, Gennadios and Schinas, is characteristic. Schinas thought that school libraries are entirely attached to the learning process and located them inside the premises of secondary schools, so that *"each student can be assisted, either studying or copying"* from their books [11]. Konstandas and Gennadios, on the other hand, admitting that *"the compilation of a library requires a great deal of funds"*, thought that it was possible to establish only one central school library that would also serve the provinces. In order to enrich it, the government should have addressed *"one or two prudent Europeans"*, who could perhaps decide to "contribute their works" to that library [11].

However, Kapodistrias was assassinated and his vision for an education system responding to the needs of the time discontinued after the first King of Greece, the Bavarian Otto (since 1833), came to the throne. Despite this, the will to establish and operate school libraries in Greece was expressed very early – almost from the beginnings of the existence of the modern Greek state and while other educational reforms were being applied under the influence of the Bavarian system. It seems that their significance in the smooth and effective operation of education, the cornerstone of the new state, had already been acknowledged. Thus, as Otto's government was particularly concerned about the education system, the first state action was taken to create school libraries, following a Royal Decree in 1835 [12], which provided for the establishment of school libraries in the country's schools *"to be used both by teachers and students"*. The same decree provided for the appointment of a person 'temporarily' responsible for each school library, a position to be occupied by the respective Principal or Head of school (Scholarch). This reference to the temporary character of this act is likely to reflect the Government's intention to address the issue of school libraries in the future too. Therefore, even this governmental intention that we notice, if we are right about rendering this temporary character correctly, implies a positive attitude on the part of the state to address the operation and management of school libraries in a systematic manner. Another possibility is that this reference was made to prevent the Principals' of Scholarch's reaction. But it those years it was unlikely for a state official or employee to react to any governmental intention expressed by law.

As for the resources of school libraries established under the aforementioned decree, they would come from the Central Library as *borrowed*, a fact which, on the one hand, was dictated by the tightness of the time, but, on the other hand, put limitations to their self-sufficiency [5]. It was provided that primary school libraries would be attended by the relevant Municipal authority. (Note that the responsibility of Local Authorities for schools and libraries, had already been officially established since 1833 under a Royal Decree issued on 26th April/8th May "Responsibilities of Prefects and Prefectural departments", which stated that Prefects would be responsible for any matters related to schools and the establishment of libraries – but did not specify whether these libraries were school or public libraries.)

The early interest in school libraries, which led to the aforementioned decree, derives, on the one hand, from the tradition of the prerevolutionary period and the period marked by Kapodistrias' governance and, on the other, by the Bavarian perception of their importance. What also played a significant role was the arrival of Greek scholars living abroad, who brought with them comprehensive ideas about the value and need of libraries in order to establish a modern national state, within the framework of which a central national library would be at the head of a system of local, special and school libraries working as tools to educate the people [13]. Moreover, the intention to create school libraries found the physical prerequisites necessary in the extensive publishing activity evident at the time and the passionate promotion that Greeks and philhellenes of the time made about offering books which made up the core of the National Library and school libraries [13]. Thus, the positive intention to create school libraries was followed by the dispatch of books to schools [14].

What followed that simple, yet pioneering for its time, Royal Decree that constitutes a primary form of intention to strengthen the institution of school libraries? What was the future of that act which was infused by the optimistic yet ungrounded mentality of "we decide and command" without measuring out the real difficulties of the time and place? "Regent Armansperg, who signs the law, used the phrase *we command* which was meant to remind people just that certain things cannot be done with orders" and nothing else [15].

The ungrounded optimism, however, that characterised the Royal Decree of 1835 was not evident in the following decree which referred to school libraries and was issued in 1855 [16]. More specifically, that decree also provided for the establishment of school libraries, but *gradually*, and only in secondary education schools. At   the same time the detailed reference to storerooms, a special register and the measures taken to protect each library's resources either from removal or damage all imply some of the difficulties already observed in previous years; moreover, they reflect a certain atmosphere and the legislator's attitude which was less creative and positive towards upgrading the functionality of school libraries with a decisive intervention[5]. This law, however,

remained inapplicable in practice, just like other laws issued in the following decades. This fact does not surprise those studying the history of Greek education, though, since the Greek educational legislation[6] is generally marked by a series of reforming efforts revoking each other and failing in practice [18].

With similar acts, which were never realised the government often expressed its intentions regarding the establishment and smooth operation of school libraries. Thus, in the 20th century, Laws 5045/1931, 5911/1933 and the Royal Decree of 15th/21st January 1949 provided that each school should establish its own school library. In the 1970's a new act (in circular 21326/20-2-1973 of the Ministry of Education [4]) stated that *"all school libraries, either in existence or under establishment, will be fully utilised…"*. All this, however, remained inapplicable while the various forms of school libraries, following the distorted development of the Greek educational system, remained inert and enclosed in dusty furniture, or, in the best circumstances, in dusty rooms, often characterised by opportunistic collections of their resources. In fact most of them worked occasionally, and only thanks to the volunteer work of people who were involved in education and believed in their educational role. It is actually known that ages ago and until the present day the books that form the resources of school libraries are found on the shelves of some (usually old) lockers in the principal's or teachers' office [19, 20] – if not in the school storeroom. However, what we should also note about the aforementioned phenomenon often occurring in Greece, of a school library responding to "a cupboard or some shelves with books" located somewhere in the school, is that there are historians of school libraries – not of Greek origin – who consider that the development of the institution of school libraries globally responds to a supposed evolution from "two shelves of books in a cupboard in the corner" to "the modern school library resource centre." W.A. King for instance, outlined a five-stage pattern of development for school libraries:

1. Small collection of books in classrooms.

2. The "storeroom" stage with books stored centrally (usually in a cupboard), and with a regular teacher in charge.

3. A library room without a librarian in charge.

4. A library room with a librarian in charge.

5. The library "programmed" as part of the teaching and

learning process in a school." [2]

Turning to the situation in Greece, we notice that the issue of school libraries was more seriously addressed   after the '80s, when school libraries were included in discussions regarding the country's educational policy.   Moreover, Art. 43 of the framework law 1566/1985 related to education provided – for yet another time in 150 years that had passed since the first legislative action in 1835 – that *"there is a school library in every school of   primary and secondary education…"* among other issues related to school libraries which were characterised as   relatively vague and unclear, mainly because they were not specialised in any later presidential decrees.

Comparing the Law of 1985 with that of 1835, it is easily ascertained that it is driven by the same air of ungrounded optimism without any realistic prospective. School libraries were not considered to be a fundamental   component of the educational policy, while their operation remained incompatible with the curriculum, which   was - and still is - prevailed by the leveling idea of using only one book for each subject [21].

Finally, the situation is changing as the 21st century is progressing. Thus, in the period between 1996-2000 a total of 499 school libraries were established in secondary education (in a total of 3.460 schools of that level)   following a balanced geographic allocation pattern around the country. The programme was funded by the 2nd Community Support Framework and national resources. The school libraries that had been officially established and could make a significant contribution to the modernisation of Greek education were appropriately rearranged   in order to make access easier and autonomous, equipped with modern material and technical infrastructure and   enriched with approximately 6.000 titles, which also included 100 titles of audiovisual resources and 10 subscriptions to magazines. It is important that a text was compiled with detailed special specifications about the area and facilities of school libraries, their resources were recorded in a special register, 49 librarians to support their operation at the central level of the Ministry of Education were hired and 450 teachers selected to recruit the above libraries on a full time basis were trained. Finally, 19 vehicles were purchased (mobile libraries) to serve remote areas. The network was expanded in 2009-10 with another 266 school libraries via the 3rd Community Support Framework (out of which 258 finally worked). Their core collection of 6.000 titles was selected by a library committee (per school unit) from a list of 13.418 approved titles [22, 23, 24, 25].

The purpose behind the development of libraries in schools, according to the statements of official bodies, was "to upgrade the education process, specifically aiming at helping teachers prepare for their teaching sessions, training students to use information resources different from the school book and thus introducing them to self- learning processes assisted also by electronic information resources (CD-ROM, Internet) and developing research and critical analysis skills" [24].

In the years when school libraries were in operation, they made a substantial contribution to Greek education as an inseparable part of the school, a centre of multimedia and information and a place where learning innovations were applied and student creativity was expressed. Running a reading room, using ICT and searching the web, sharing classes between teachers, supporting the compilation of projects and references, participating in innovative initiatives – such as cultural programmes, health education programmes, e-twinning programmes, etc.-, programmes promoting reading and also reading clubs, organising events (such as visits of writers and scientists, music events, book exhibitions, theatrical performances, film/video projections, painting competitions, presentations of museum boxes, issuing school magazines, one-day seminars), are some of the activities developed through school libraries [26].

Adverse conditions, however, once again did not allow school libraries to flourish. And that was not due to general difficulties related to their operation [27], which could perhaps be surmounted, but to the economic recession which led school libraries to partial operation. More specifically, since 2010-11 208 of a total of 757 school libraries did not work because teachers were not transferred to them to serve as those responsible for their operation. Even worse in the school year 2011-12, following a circular of the Ministry of Education (750/Φ 2.2/2-9-11) it was dictated that their equipment and technical infrastructure be returned and the teachers responsible for their operation – specially trained and experienced – were called upon to return their keys and go back to their original teaching positions. On the contrary, certain teachers were appointed responsible for the operation of school libraries in order to complete their teaching timetable, without any training requirement or even desire to do so; just to complete their teaching timetable. Moreover, they spent only that many teaching hours (i.e. 40-minute hours) as were necessary for them to complete their teaching timetable – e.g. only 4 teaching hours per week. Therefore, in the school year 2011-12 school libraries remained open for 2 hours 3 or 2 times a week; that is if they opened. So it goes without saying and is obvious that they did not work regularly and their operation was once again bestowed to the volunteer offer of work of the people who believe in them. Realising, however, the important role that the personal character and strength or personality of people working in school libraries and also those who are somehow involved with them has played in the development of an institution such as school libraries, especially when the institutional framework is so poor and the applied state/governmental policy about school libraries is either nonexistent or inadequate, is a fact ascertained also in the international published works. In Marquardt, there is a characteristic reference to the personal character and the informal networks created by those involved in school libraries: "The several goals which are outlined in the IFLA-UNESCO Manifesto [about school libraries] are usually met only in part, due to the lack of different kind of resources (e.g., collections, space, equipment and staff). Nevertheless there is a progressively increasing number of best practices which are stimulated by formal or informal networks, national or local projects, and above all, the commitment and the enthusiasm of the practitioners themselves in the field." [28]

Of course the fact of the aforementioned policy of the Ministry about school libraries and their staff was commented sharply and bitterly in the press by those responsible for school libraries, organizations, bodies, etc.; a fact which shows how sensitised the public and those involved in education and libraries in Greece are towards school libraries. Here are some characteristic headlines: *School libraries to be closed / School libraries, time for inactivation / Do we need school libraries? / All 700 school libraries to be shut down / School libraries are now being closed - A cry of anxiety from those responsible for their operation / School libraries to be abolished / God willing!* and others [29].

CONCLUSION

If the present of school libraries is so uncertain, though, what about their future? However, it is commonly held that a modern state should receive good education, which is even more necessary in a period of crisis. And it is commonly agreed that school libraries are necessary to support that education.

According to statements made by the Greek Ministry of Education itself: *"the institution of school libraries supports education and enhances learning in a way that is innovative for Greek standards, beyond the traditional teaching framework. The individual is encouraged not to content him/herself with traditional books, but seek information also in other sources. In the modern society of knowledge and information it is particularly important that students are familiarised with the information process in every stage of their everyday lives, not only as part of education. As time progresses, school libraries are developing into an essential part of a new educational reality, moving away from the traditional model of a lending school library and acquiring a new important role, which is complementing and supporting to classroom teaching."* [25]

And also elsewhere: *"School libraries were established in order to be used as:*

- *(for students) a place to learn, be informed, be acquainted and familiar with books and the information society.*

- *(for teachers) a source of information and a tool suitable to design new learning methodologies.*

(for the educational system) a channel to promote an educational school environment which is changing and reformed syllabi through the integration of school libraries. What remains to be achieved is the realisation and implementation of the role of school libraries as educational tools incorporated into changing educational approaches and distinguished as one of the most effective educational tools" [30]. It is true that numerous studies conducted in many countries have proved that there is a positive relation between school libraries and school achievement - and even personal development [31] –of students who use them [32][7]– especially when the people responsible for them are trained and collaborate productively and constructively with the teachers of the schools they support.

Let's sum up, then, with a particularly well-aimed extract: "Will our children be ready for the challenges of their future? Are our schools ready to prepare them for those challenges today? *Center for International Scholarship in School Libraries* takes the position that schools without school libraries cannot educate this generation in a way that prepares them for 21st century study and work, and being part of the increasingly digital, global society. Cutting school libraries is not the solution: School libraries, now more than ever, are integral to quality learning and teaching in 21st century schools." [34] Thus, we return of the prerevolutionary yet timeless urge of Adamantios Korais *"Seize Learning"*, supplemented by other bitter-sweet ascertainment of the same thinker: *"Educating peoples is a time-consuming task"* [35].

REFERENCES

1. Clyde, L. A. (1981). The magic casements: a survey of school library history from the eighth to the twentieth century. PhD thesis, James Cook University http://eprints.jcu.edu.au/2051/ (recovered on 13-8-12)

2. Clyde, L. A. (2002). The Schole Lybrarie: Images From Our Past http://www.iasl-   online.org/my/search/?s=Laurel+A.+Clyde (recovered on 13-8-12)

3. Muller, K. (2011). First School Library? http://www.americanlibrariesmagazine.org/ask-ala-librarian/first-school- library (recovered on 13-8-12)

4. Kuić, I. (2009). School reform during the second Austrian administration in Dalmatia, and primary school librarieshttp://hrcak.srce.hr/index.php?show=clanak&id_clanak _jezik=58387 (recovered on 12-8-12)

5. Papadopoulos, D. (2012). Evolution of legislation of school libraries from 19th century until now and its relation to Greek reality, dissertation project, TEI of Athens, department of Library Science and Information Systems [in Greek]

6. Dalakoura, E., Bota, D. (2009). Secondary education libraries at Etoloakarnania: History and law framework, dissertation project, TEI of Thessaloniki, department of Library Science and Information Systems [in Greek]

7. Papazoglou, A. (1998). School Libraries in Greece: A State-of-the-Art Report http://www.iasl-online.org/pubs/slw/slwjuly98.html (recovered on 12-8-12)

8. Bikos, G. D. (2011a). Research on School Libraries in Greece and Suggestions on its Further Development. http://pubs.i-das.org/aipm/index.php?option=com_k2&view=item&id=16:p roceedings-icininfo-2011&Itemid=61 (recovered on 22-8-12)

9. Bikos, G. D (2011b). Research field of school libraries: Registration of its data and its perspectives. Libraries & Information, 23, 32-37. [in Greek]

10. Kokkinis, S. (1974). Elementary school libraries in Greece. Athens [in Greek]

11. Geledaki, S.. (1998). Provision of knowledge during the Kapodistrian era. Curricula and textbooks. Doctorate thesis. Athens: issued by University of Athens [in Greek]

12. Royal Decree, 8(20)/11/1835, official gazette 20/16-12-1835: "About structuring public libraries in every state school." [in Greek]

13. Sklavenitis, T. (1989). The library of University of Athens (19th century), In Minutes of international symposium (vol. A´), University: Ideology and Paideia, 113-119, Αθήνα [in Greek]

14. Declaration of the Secreteriat of Ecclesiastics and Public Education, 29/11/1835, official gazette 21/22-12-1835: About the type and quantity of supplied books by the state to public schools." [in Greek]

15. Delopoulos, K. (1989). "At every primary and secondary school operates a school library." A law from 1835 that has almost never changed after 154 years... Routes in children's literature, 13, 10-15. [in Greek]

16. Royal Decree, 28/10/1855, official gazette 46/3-12 1855: "About distribution of bought or donated books." [in Greek]

17. The Ministry of Ecclesiastics and Public Education, Athens, 28/12/1856: Specifying the books stored [in Greek]

18. Dimaras, A. (2007). The transformation that has not happened, History documents: 1821-1894 vol. A´ and History documents: 1895- 1967, vol. B´. Athens: Estia Bookstore [in Greek]

19. Saiti, A. X. (2002). Educational reality of primary schools of Etoloakarnania ans Attiki through the operation of school libraries. Athens: Atrapos. [in Greek]

20. Arvaniti, K., I. (2008). The library in Greek school. The institution, the reality and its ideal form according to teachers. Thessaloniki: Ant. Stamouli [in Greek]

21. Arvaniti, I., Kyridis, A., Dinas, K. (2007). Greek Primary School Teachers Dream of the Ideal School Library, Library Philosophy and Practice http://www.webpages.uidaho.edu/~mbolin/ arviniti-kyridis-dinas.htm (recovered on 14-8-12)

22. Arahova, A. & Kapidakis. S., Empowering our Libraries, Empowering our Education System: Using research results to shape policies that optimize the utility of school libraries http://www.ionio.gr/~sarantos/repository/c31C-Bergen.pdf (recovered on 12-8-12)

23. Arvanda, A. New school libraries http://1lyk-elefs.att.sch.gr/library/arvanta_nees_viv.pdf (recovered on 14-8-12) [in Greek]

24. http://www.epeaek.gr/epeaek/el/a_2_2_2_2_1.html (recovered on 13-8-12)

25. http://www.ekt.gr/school-library/project.htm (recovered on 11-8-12)

26. Ninos, E. (2011). The role of school library, Working meeting regarding school libraries at 3rd General Lyceum of Ilioupoli http://www.box.net/shared/biddqarngp (recovered on 11-8-12) [in Greek]

27. Barberis, G. (2011). Issues posing difficulties at school libraries operation, Working meeting regarding school libraries at 3rd General Lyceum of Ilioupoli

http://www.box.net/shared/rdyygop5d1 (recovered on 22-7-12). [in Greek]

28. Marquardt, L. (2008). The Leopard's Spots on the Move: School Libraries in Europe, reported submitted to IFLA http://eprints.rclis.org/bitstream/10760/14272/1/marquardt_fi nal4IFLA_20080818.pdf (recovered on 12-8-12)

29. http://greekschoolibs.blogspot.gr (recovered on 22-7-12)

30. http://archive.minedu.gov.gr/el_ec_page2101.htm (recovered on 14/8/12)

31. Fodale, F. & Bates, J. (2011). What Is the Impact of the School Library on Pupils' Personal Development? A Case Study of a Secondary School in Northern Ireland http://www.highbeam.com/doc/1P3-2510641661.html (recovered on 13-8-12)

32. http://www.iasl-online.org/advocacy/make-a-difference.html (recovered on 13-8-12)

33. http://www.lrs.org/impact.php (recovered on 21-7-12)

34. Todd, R. J. & Kuhlthau, C. A. (2005). Student Learning Through Ohio School Libraries, Part 1: How Effective School Libraries Help Students, School Libraries Worldwide, Volume 11, Number 1, 63-88.

35. Velegraki, A. (2005). Korai's up-todate pedagogical beliefs http://www.patris.gr/articles/59249?PHPSESSID (recovered on 14-8-2012) [in Greek]

Chapter 3

Redefining Library Learning Facilities in Malaysia: Lesson from Frank Lloyd Wright Sustainable Approach in Spatial and Landscape Design

N. Utaberta [a] *, N.Spalie [a], MM.Tahir [a] , NAG.Abdullah [a]

[a]University Kebangsaan Malaysia,Department of Architecture,Bangi 43600,Malaysia

ABSTRACT

The main objective of this paper is to initiate and open a wider discussion on library design and learning facilities in Malaysia. It tries to take some lesson from one of the greatest modern American Architect which is Frank Lloyd Wright. The discussion itself will consist of four (4) parts. First part will discuss on the current problem and situation in reading and library issues in Malaysia while the second part will try to explore and review on the definition of outdoor learning with its importance and character in learning environment. The third part will focused on the Frank Lloyd Wright's sustainable approaches in spatial and landscape arrangement while the last part will do some analysis and comparison which result on the suggestion and recommendation for outdoor learning facilities design in Malaysia. With some understanding from the established environmental approaches of Frank Lloyd Wright's above it is expected that we can reconstruct and redefine some framework and guideline for future outdoor facilities design in Malaysia.

Keywords: Outdoor Library Learning Facilities Design, Frank Lloyd Wright, spatial and Landscape arrangement;

INTRODUCTION

In 1982, the National Literacy Survey carried out by the National Library reported that Malaysians only read an average of one to two *pages* a year. Fortunately, the reading habit among Malaysians improved to two books per year when the National Literacy Survey was repeated in 1996. Nonetheless, the last National Literacy Survey carried out in 2005 reported that Malaysians still read an average of two books a year. In short, there had been no improvement.

The last survey also reported that Malaysians read increasingly less as they grew older. By the age of 50, for example, only 20% of Malaysians would still continue to read books, a drop from 40% (a figure which is already pathetic) from those in the mid-twenties to thirties age group but survey also stated that the children in Malaysia read less book than the adult.

READING AND LIBRARY ISSUE IN MALAYSIA

Although Malaysia has among the highest literacy rates in South-East Asia, Malaysians prefer light reading material like newspapers and magazines to books. Deputy Prime Minister and Education Minister Tan Sri Muhyiddin Yassin said at the launch of the Kuala Lumpur International Book Fair 2009 at the Putra World Trade Centre. According to literacy statistics, out of 85% of Malaysians who read regularly, 77% of them prefer newspapers, 3% read magazines, 3% read books and 1.6% read comics.

"If we were to compare with American citizens, 53% of them read fiction and 43% of them read non-fiction books." Malaysians are more inclined to read light material while citizens from developed countries read books,The UN Development Programme's 2007/2008 report said the literacy rate of Malaysians, at 93.2%, was behind developed countries like Japan, Britain, the United States, Australia and Germany, which have literacy rates of 99%. We still have room for improvement in terms of increasing the literacy rate to 99% by 2020 and the quality of material that we read. Parents play an important role in nurturing the reading habit among their children. Students should read more books, not just revision books or textbooks for examinations."

Prof. Ambigapathy Pandian from Universiti Sains Malaysia (USM) has perhaps studied the most on the reading habits of Malaysians. In an interesting paper by him in 2000, he surveyed that 80.1% of university students are "reluctant" readers in English-language materials. In other words, 80.1% university students read because they have to. Interestingly, Malay and Indian students have a higher tendency to seek English-language reading materials than the Chinese. Based on his survey, Prof. Pandian also outlined a profile of a habitual reader in English.

The Malaysian education system is in dire straits. With the education system reverting back to Malay language as the medium of instruction in schools and the government desperately plugging all holes in a sinking boat, I strongly believe the key to improving our education is the inculcation of a strong reading habit among all Malaysians. Although the government has launched several reading campaigns (the recent one is the *Mari Membaca 1Malaysia*, launched in March 2010) to increase the reading habit among Malaysians throughout the years, obviously these campaigns aren't quite working as desired.

A reading habit is an essential life skill. Reading not only increases our knowledge, but it also builds maturity and character, sharpens our thinking, and widens our awareness in social, economic, political, and environmental issues. What most of us don't know that, unlike speech, reading is a learned skill; our brains aren't hard-wired to read. Although a baby can pick up speech from listening to others talking, reading requires learning. In other words, reading takes effort. It is hard work. But it builds our brain muscles. The effort to inculcate a reading habit pays off handsomely, either directly or indirectly, in our lives.

CHILDREN LEARNING DEVELOPMENT AND THE IMPORTANCE OF OUTDOOR LEARNING

The National Association for the Education of Young Children (NAEYC) has articulated a position statement on principles of child development and learning that inform developmentally appropriate practice (NAEYC, 1997). Child development encompasses several domains: cognitive, social, physical, and emotional. While developmental stages occur in an ordered progression and new skills are dependent on old skills, the rate at which each child develops differs. In addition, an individual child can progress through different domains at different rates (NAEYC, 1997). That is, not only do different children develop at different rates, but an individual child may progress unevenly within different domains. For example, a child may have a highly developed sense of kinetics, spatial relationships, and fine motor skills. The same child may experience language difficulty. Children at the same developmental level may have different ways of knowing and learning and different ways of demonstrating what they know (Vandergrift, 1996; Gardner, 1999). Each child is unique and "variation is not only to be expected but also valued" (NAEYC, 1997). Children's skill in spelling, typing, spacing, punctuation, syntax, alphabetization, scanning, and tracking may vary (Busey & Doerr, 1993). Children in the same class at school may differ in their ability to decode, follow directions, and stay on task. In order to accommodate varying developmental rates, learning styles, and preferences of children in the same class who share hardware and software, systems should have, for example, the option of keyboarding or point and click navigation.

Children need to broaden and deepen the knowledge they already have, and they need the opportunity to relate this new information to something in their experience that they already understand (NAEYC, 1997). They need both the challenge of new experience and the opportunity to practice skills they already possess. Vygotsky's Zone of Proximal Development is "the distance between the actual developmental level as determined by independent problem solving and the level of potential development as determined through problem solving under adult guidance or in collaboration with more capable peers" (Vygotsky, 1978, p. 86). A variation of this is Kuhlthau's Zone of Intervention, "that area in which a user can do with guidance and assistance what he or she could not do alone" in the Information Search Process (Kuhlthau, 1993, p. 176). Developmentally appropriate outdoor and natural environments for children support both mastery of knowledge and growth. Play is an important part of a child's social, emotional, physical, and cognitive development. It gives the child an opportunity to practice new skills and construct meaning without risk (NAEYC, 1997). Play is important because it enables children to become familiar with materials and concepts. Play becomes even more valuable when it offers feedback that the child can interpret (Bowman & Beyer, 1994).

There are many definition in order to express the meaning of the outdoor learning. The definition are classified into two(2) categories, *'Psychosocial definitions'* and *'Environmental Definitions'*. Based on C. A. Lewis, 1975, The Administration of Outdoor Education Programs. Dubuque, IA: Kendall-Hunt, the outdoor learning in term of *'Psychosocial definitions'* is- *"appeals to the use of the senses - audio, visual, taste, touch, and smell - for observation and perception."*. For the other side definition- *'Environmental Definitions'*, the outdoor learning have been defined as - *"an experiential method of learning by doing, which takes place primarily through exposure to the out-of-doors. In outdoor education, the emphasis for the subject of learning is placed on RELATIONSHIPS: relationships concerning human and natural resources."* Priest, S. (1988). Another definition can be find as:

-*"an experiential method of learning by doing, which takes place primarily through exposure to the out-of- doors. In outdoor education, the emphasis for the subject of learning is placed on RELATIONSHIPS: relationships concerning human and natural resources."* Priest, S. (1988).

Based on the previous discussion about the development of children's teaching and learning its clearly shown that each of the child required a conducive environment to grow and learn, is make outdoor learning will be one of the importance element in developing their reading and intellectual interest. This is a stepping stone for a new preschool's teaching and learning system and module in Malaysia in order to produce more intellectual, critical, outspoken, independence and expression generation.

STUDY ON FRANK LLOYD WRIGHT'S THOUGHT IN LEARNING FROM NATURE

Frank Lloyd Wright was one of the best references of an architect whose design is very close with its nature. The building is always designed to be part of its landscape setting. Even though not many record stated about Mr. Wright designing a library but his approaches and understanding of nature will be a valuable reference for the outdoor learning setting and design of library facilities in Malaysia. Based on the understanding, nature was "The only body of God that you can see", Frank Lloyd Wright strongly hold the concept of nature as the reflected God that should respected and appreciated by man.

"The real body of our universe is spiritualities-the real body of the real life we live. From the waist up we're spiritual at least. Our true humanity

begins from the belt up, doesn't it? There in comes the difference between the animal and the man. Man is chiefly animal until he makes something of himself in the life of the spirit so that he becomes spiritually inspired-spiritually aware. Until then he is not creative. He can't be. "(Wright, Frank Lloyd, Truth against the World, pg 270)

This clearly shown from the prairie houses and most of his public building a consistency that a building is a man's product that should bow and adapt their own character and strength from the nature and its natural surroundings. Building should able to adapt with its context. This approaches made Frank Lloyd Wright buildings integrated with the site, does not shouting arrogantly to show themselves but speaks harmony with anything that be around them.

Figure 1. Frank Lloyd Wright buildings that shown the power and respect to the nature and surrounding to the place of each building that has been constructed.

The result of Frank Lloyd Wright life's research that his original thought and concept comes from his childhood environment. Attention from his mother and his experience such along working in his uncle's farm has formed and practice his mind to appreciate and look into the nature as an element that cannot be separated in once design planning. That is the main factor to determine a design.

"She (Wright's mothers) loved to pick windflowers in the hills and meadows, studying them, arranging them in cluster, explaining to him the intricate formation of the petals in relation to leaves and stem. She love

ferns because of their geometric design and passed that love to her son…" (Wright, Olgivanna Lloyd (1966), Frank Lloyd wright: His Life, His Work and His Words, pg 15).

Idea and thinking that the nature owns a great power have a large and important implication in design. This make Wright's architecture speak through the material utilization, design process and how to standardize the scale and comparison in a very nature and humane ways. The idea of the power of the nature has also produce a concept about sustainable life. There are two dimensions in Wright thinking and philosophy in sustainable life: nature dimension as physical aspect and social dimension which is thinking and man understanding as spiritual aspect. The nature dimension talks about the struggle to look after and treat the nature as a God production that should be sustained, even the quality of supportive side should be upgraded while the social dimension brought the idea of social responsibilities and community understanding.

"We must conceive and integrate: begin again at the beginning to build the right kind of building in the right way in the right place for the right kind of people." (Wright, Frank Lloyd (1949), Genius and Mobocracy, pg 13)

Wright starting point learning as an architect is comes from Anna Lloyd Jones, his own mother which is a teacher and his main inspiration. However his mother famously known as an influential and intelligent women with her vision and strong desire. This clearly shown in Wright writing which is explaining the perception of his mother environment.

"Education was Sister Anna's passion even while very young. All this family was imbued with the idea of education as salvation…soon she became a teacher in the countryside, riding a horse over the hills and

through the woods to and from her school each day. Old men in the neighborhood still speak of Sister Anna as their teacher, with admiration and respect."(Wright, Frank Lloyd (1943), an autobiography, pg 9).

"Mother learned that Frederick Froebel taught that children should not be allowed to draw from casual appearances of nature until they had first mastered the basic form lying hidden behind appearances" (Wright, Frank Lloyd (1957), a Testament, pg 19).

The education of this basic shape trained the Wright's ability during his childhood until he become to be a great designer *"All in my fingers to this day"*, the confession of Wright. With this method in Wright opinion, designing was fun. Other source of Wright's design talent is came from "German Paper" with interesting colour , usually it was cut and will be arranged suitable with his desire, structural game with toothpick, straw and plant and dry twig. Even, Anna always asks Wright to look into the nature and teaches a lot of things to Wright.

Wright's understanding and his love to the nature keep growing when he was send by his mother to stay with his uncle, James Lloyd at Wisconsin. In these adventures alone-abroad in the wooded hills to fetch the cows, he, barefoot, bare head urchin, was insatiable, curious and venturesome. So he learned to know the wood, from the trees above to shrubs below a grass beneath. And the millions of curious lives had hidden in surface of the ground, among roots, stems, and mold. For Wright, to form a building must begin from the natures. Build a structure with nature integration and human without damaging the natures surrounding.

APPLICATION OF WRIGHT APPROACHES IN LIBRARY LEARNING FACILITIES

As the discussion above, Wright's approaches to the nature has produced a thought in the application of the concept in outdoor learning facilities. There are several examples of nature approaches in the outdoor learning school that so called Sekolah Alam in Indonesia as can be see below:

Figure 2. Some outdoor learning facilities as a tool to develop children's interest in reading the nature

This nature approach has means with the Wright words:

"Why is any cow, red, black or white, always in just the right place for a picture in any landscape? Like a cypress tree in Italy, she is never wrongly placed. Her outlines quite down so well into whatever contours surround her…"(Wright, Frank Lloyd. An Autobiography. pg 23).

Facility design that respect nature without losing the nature structure that certainly formed naturally because to Wright the natures itself is *"The only body of God that you can see"*. Nature beneficial in learning system also is one adaptation of Wright approach that respects nature because there are thousand answers to all issues as he stated:

"Young Wright saw that nature was a wonderful teacher and had answers to many question that theoretical learning could not explain nearly so well." (Peter Blake, Master Builders, pg 270).

For Wright, to form a building must begin from the natures. Build a structure with nature integration and human without damaging the natures surrounding is in his heart.

CONCLUSION

A lesson from Wright's design in outdoor learning design application has produced a learning process- *"Smart Green Education"*. Another new method with the concept of respect and utilize the nature as a main method in teaching and learning for Pre-School stage. It is a experience learning method by doing and practical in exposing the outdoor world. In outdoor education, the core of the learning is place on RELATIONSHIP: a relationship about man and nature sources. Facility design that respect nature without losing the nature structure that certainly formed naturally because to Wright the natures itself is *"The only body of God that you can see"*.

REFERENCES

1. Blake, Peter (1960). *The Master Builders*. New York: Alfred A Knopf.

2. Bowman, B., & Beyer, E. (1994). Thoughts on technology and early childhood education. In J. L. Wright & D. D. Shade (Eds.), Young children: Active learners in a technological age (pp. 19-30). Washington, DC: National Association for the Education of Young Children.

3. Busey, P., & Doerr, T. (1993). Kid's catalog: An information retrieval system for children. Journal of Youth Services in Libraries, 7(1), 77-84. Curtis, William JR (1982). *Modern Architecture since 1900*. Oxford: Phaidon Press Ltd.

4. Gardner, H. (1999). Frames of mind: The theory of multiple intelligences. New York: Basic Books. Heinz, Thomas A (1996). *Frank Lloyd Wright: Field Study*. London: Academy Editions.

5. Heinz, Thomas A (2002). *The Life and Works of Frank Lloyd Wright*. Kent: Grange Books Plc. Hitchcock, Henry Russell (1941). *The Nature of Materials*. New York: Da Capo Press, Inc.

6. Hoffman, Donald (1978). Frank lloyd Wright's Fallingwater: The House and Its History, New York: Dover Publication Inc. Kaufmann, Edgar J (1989). *9 Commentaries on Frank Lloyd Wright*. Massachusetts: MIT Press

7. Kuhlthau, C. (1988). Meeting the information needs of children and young adults: Basing library media programs on developmental states.Journal of Youth Services in Libraries, 2(1), 51-57.

8. Laseau. Paul (1937). *Frank Lloyd Wright; Between Principle and Form*. New York: Van Nosrand Reinhold.

9. National Association for the Education of Young Children (NAEYC). (1996). Technology and young children--Ages 3 through 8 [Position statement]. Washington, DC: NAEYC. Retrieved October 12, 2005, from http://www.naeyc.org/about/positions/PSTECH98.asp.

10. National Association for the Education of Young Children (NAEYC). (1997). Principles of child development and learning that inform developmentally appropriate practice [Position statement]. Washington, DC: NAEYC. Retrieved October 12, 2005, from http://www.naeyc .org/about/positions/dap3.asp.

11. Nute, Kevin (1993). *Frank Lloyd Wright and Japan*. New York: Van Nosrand Reinhold.

12. Pandian, A. (2000). A study on readership behaviour among multi-ethnic, multi-lingual Malaysian students. A paper presented at the seventh International Literacy and Education Research Network (LERN) Conference on Learning, RMIT University, Melbourne, 5-9 July 2000.

13. Perry, Marvin (1981). *Western Civilization*. Boston: Houghton Mifflin Company. Peter, John (1958). *Masters of Modern Architecture*. New York: George Braziller Inc.

14. Pevsner, Nikolaus (1936). *Pioneers of Modern Architecture*. Middlesex: Penguin Books. Pevsner, Nikolaus (1943). *An Outline of European Architecture*. Middlesex: Penguin Books. Peel, Lucy (1989). *An Introduction to 20th Century Architecture*. London: Quantum Books Ltd.

15. Pfeiffer, Bruce Brooks (1984). *Letters to Architect; Frank Lloyd Wright*. California: California State University Press. Vandergrift, K. (Ed.) (1996). Ways of knowing: Literature and the intellectual life of children. Lanham, MD: Scarecrow Press.

16. Vygotsky, L. (1978). Mind in society: The development of higher psychological processes. Cambridge, MA: Harvard University Press. Wright, Frank Lloyd (1943). *An autobiography by Frank Lloyd Wright*. New York: The Frank Lloyd Wright Foundation.

17. Wright, Frank Lloyd (1949). *Genius and Mobocracy*. New York: Horizon Press.

18. Wright, Frank Lloyd (1957). *Truth Against the World. New York: A Wiley-interscience Publication.*

19. Wright, Frank Lloyd (1957). *A Testament*. London: Architectural Press. Wright, Frank Lloyd (1958). *The Living City*. New York: Horizon Press.

20. Wright, Olgivanna Lloyd(1966). *Frank Lloyd Wright; His Life, His Work, His Words*. London: Pitman Publishing. Wright, Frank Lloyd (1954). *The Natural House*. New York: Horizon Press.

Chapter 4

'In-formation' of Better Learning Environments - the Educational Role of the University Library

Thomas Hapke

ABSTRACT

This paper gives an overview of the close connection between learning - which is seen as a constructivist, active, self- directed and social process - and the university library. It discusses the role of the university library in the electronic learning environment, and considers the physical library as a place of learning which is not out of date. It also considers information literacy as the key to the library's educational role regarding its contents. Organisational issues hold together the three points mentioned above, the learning library as the prerequisite of the learning-empowering and learning-facilitating library.

INTRODUCTION

The educational role of the library is a crucial one for the future, not only because of the etymological roots of 'in- formation' in instruction and education. Furthermore this role cannot be seen only in regard to e-learning. Libraries and librarians, with their core competencies, especially in providing advice and curation, are important partners in the creation of better learning environments.

To begin with, a library in itself is educational. Learning for the present and the future is only possible when using knowledge and experiences from the past. Without the awareness of the past, planning for the future is like trying to plant cut flowers. [1] The records from the past and its heritage can be found in memory institutions like archives, libraries and museums. So the use of libraries is a natural part of education.

This traditional role of libraries is still emphasized by academic libraries, e.g. in some traditional advertisements at the library of the University of Illinois at Urbana-Champaign - one from 1926, the other from today.

Picture 1:

To see the library in close connection to education and learning is also a very old view, as the following citation proves (Dury, 1650, p.17-18): "...Librarie-keepers ... ought to become Agents for the advancement of universal Learning ... his work then is to bee a Factor and Trader for helps to Learning ... " About 250 years later William Poole emphasized the importance of how to use books as important part of education, when he wrote: "the study ... of the scientific methods of using books should have an assured place in the University Curriculum. ... [it] will aid them [the graduates] in their studies through life." (Poole, 1894, p.3)

Like a previous paper by Virkus and Metsar, this paper also gives an overview of the close connection between learning and the university library today and in the future (Virkus & Metsar, 2004). While it speaks of the role of the university library as a service institution it also considers the role of the university itself, and here especially about the future of learning in general. Learning has changed and will change. But e-learning is only one part of it. Libraries' services have to change in response to changes in universities' mission and programs.

LEARNING AND LEARNERS IN THE UNIVERSITY LIBRARY TODAY

Remember a situation in which you really have learned something. For the author of this paper an example would be preparing this presentation for the LIBER Conference 2005 in Groningen. Shneiderman describes the circle of learning as collecting, relating, creating and donating. Gathering and reading documents, then extracting information from required resources are followed by relating them to one another or to one's own experiences. Good learning processes include the connection and communication of learners, which occur best when working in collaborative teams. In an optimal situation, at the end of contemporary learning there is a product to be created - such as this paper - which should be relevant not only for the learner but also for a community outside the specific learning situation, in this case for you as a reader of this paper. The process of learning begins by dealing with information, so informing is part of every learning process and learning requires competence to find and use information (Shneiderman, 2002).

From a constructivist view, the best learning is active and self directed, including one's individual background and one's own interpretation. It is context-specific and situative as well as a social process that can create communities. Learning includes perceiving and interpreting as well as understanding and acting. In the future the transformation of subject knowledge will probably be less important than the promotion of awareness and intelligence as well as competencies and creativity. Self-observation, learning to learn, supporting the search for the "right" questions, reflecting and improving one's own competencies, research skills and knowledge and coping with the information jungle are all necessary.

The growing importance of the promotion of competencies and soft skills also influences the learning environments in universities. These can be characterised as "information ecologies" (O'Day & Nardi, 2003). In universities, as in every ecosystem, change, diversity and locality are key characteristics of teaching and learning.

Learning will become more problem- and project-oriented, including multiple contexts and perspectives. Cooperation in groups and over-the-shoulder-learning (Twidale & Ruhleder, 2004) as well as "informal" learning (Tully, 2004) will be normal. The difference between education and research will continue to diminish.

Educational institutions, and thus also libraries, have to cope with the postmodern attitudes of their keystone species, the students. Knowledge becomes fragmented. Our customers show consumer behavior and flippancy on the one side (Harley, Dreger & Knobloch, 2001), are interactive, social, and always connected. Some attend universities part-time, and many may also have families. The papers in Oblinger and Oblinger give a good overview of the challenges of educating students today and also consider the library environment. Cultural values shift from linearity to multidimensionality, from stability to continuous change, from individualism to collaboration (Oblinger & Oblinger, 2005).

The current state of using electronic subject-specific information sources was described in the report of the project "Stefi - Studieren mit elektronischen Fachinformationen (Studying with electronic subject-specific information)", which was funded by the German government (Klatt et al., 2001). Results of the Stefi-Project show that the use of electronic, subject-specific information sources is not yet sustainably implemented in study courses. Compared to their use of electronic information media, the information literacy of students is relatively low. Searching for information is too often limited to 'browsing' with the help of free search engines. Students have to acquire knowledge of how to use electronic information autodidactically, a situation which is also true for quite a few lecturers. The majority of lecturers feel that they need further education in using digital information and they think that it should become a subject of introductory and higher classes for students.

The view described above of learning and its environment serves as a basis for the following arguments in respect to the services libraries can offer to support learning.

SERVING ENVIRONMENTS FOR LEARNING THROUGH THE LIBRARY

"Today, instruction ought to be arranged in such a way that it makes it possible for the individual to change according one's own decision. But this is possible on one condition only, that teaching is a possibility offered permanently." (Foucault, 1999, p.20, own translation) This quotation shows the way to future development in education. This paper discusses four points, which are to be considered when viewing the services of libraries within the learning environment of today and tomorrow:

1. Learning and the library in the electronic environment.

2. The physical library as a place of learning which is not out of

date.

3.   The educational role of the library regarding contents where information literacy is the key.

4.   The organisational issues, which hold the previous, three points together.

(E-)learning and digital libraries

The last few years have shown a great movement towards e-learning but, as the Australian Neil McLean states, it is still in the "cottage industry phase" (McLean, 2004). On university level there is a need for a single-step approach for students to gain access to all relevant information and materials that they need for their specific lectures or courses. Ideally, students need to find what they want at one place on the Net: the hours and location of their courses, class notes, textbooks, exercises, link collections, (electronic) reserves, real e-learning material etc. Learning management systems (LMS) exist here as solutions, that grow in parallel to other electronic repositories in the university. The output of computer-supported learning consists of numerous documents, text, audio or video files, which may exist in different repositories or digital libraries. These digital libraries also contain learning objects as structured electronic resources containing contents of high quality, clear learning goals and a dedicated audience. Thus, learning objects are composites which carry - together with the object itself - the context, in which the object can be used, and which is described in its metadata. To offer such systems is a question of information management and therefore a possibility for the university library to do what the library also did in the past. The competencies of library staff are helpful here to manage them (Lynch, 2002).

It is important to make the library visible in this environment through integrating library services and LMS (McLean & Sander, 2003; McLean & Lynch, 2004). In every LMS or virtual learning environment there exist modules like chat and bulletin boards to encourage exchange between students. It is also important to facilitate the creation of information products by the patrons themselves, e.g. by creating services for digital consulting (intellectual property) and services for digital production. What is needed are places for learner expression, e.g. electronic portfolios, a form of learning diaries, whose importance were emphasized by Roes (Roes, 2001). Wikis and weblogs, as so-called 'social software', may be additional instruments and through their use in education e-portfolios may be realized. Libraries have a chance to use these new tools as communication instruments between their patrons or between the library and its users.

Many enhancements of traditional services through the library can be used to support e-learning. In the management of e-resources, metadata are a core responsibility of libraries. Virtual and real reference can support individual learning. Reserves are types of early learning management systems, which now in their electronic form have to be integrated in LMS. Until now e.g. the university library of the Hamburg University of Technology (TUHH) has only offered electronic lists of books in the 'real' reserves, sometimes with additional collections of links. Exploring services for mobile devices, representing services and instruments in a graphical and visually supported way or simply the lending of USB sticks, iPods, handhelds, laptops etc. are further activities to engage patrons where they really live and learn. Digitization and copyright clearance services for electronic reserves are especially important for faculty. An increased number of services for particular target groups (faculty, students, specific departments, etc.) are necessary. For example the TUHH library offers special web pages as subject-specific gateways into library resources, e.g. for the Northern Institute of Technology or the Hamburg School of Logistics which are two innovative public private partnerships with the TUHH.

Libraries have to offer multiple means of access for many different kinds of people. Users have to cope with a growing diversity and complexity of information resources, normally in the form of databases. Libraries undertake many different strategies to support their users in this effort: incorporating their data and holdings in Google, combining databases in federated search systems and portals as well as trying to increase reflection and awareness with information literacy activities when dealing with the information jungle (see part 3.3.).

The library as a physical learning place

Let us observe what happens in the physical library. User activities like information seeking, teaching and learning, recreation, connection and contemplation are only a few of the most important 'happenings' within a university library. The library is a place of interaction and experiencing instead of a warehouse as it was viewed in the past. It should be a place of life, a learning laboratory. [2]

There can never be enough study space in a library. Even in engineering libraries there is a need for a lot of workplaces (Picture 2). The Grainger Engineering Library Information Center at the University of Illinois in Urbana- Champaign for example has more than 700 places at tables, more than 450 in carrels, more than 100 so-called 'informal' seats such as armchairs and sofas, and more than 150 at computer workstations, but fewer than 40 catalogue and research workplaces. So, current attempts in Germany to reduce the key figure for workplaces when planning library spaces in the future, especially in engineering, should be questioned. Educating and learning is a social event, and therefore does not only happen at home!

Picture 2: *Reading room at the Grainger Engineering Library at Urbana-Champaign*

In library planning in the USA, there is a trend away from individual study carrels to table-and-chair ensembles. The traditional library reading room has been revived as a place for working privately in the presence of others - a place to see and to be seen (Demas, 2005). Reading rooms in many libraries in the States quite often look similar to normal living rooms, also containing possibilities for 'non-library' uses like meeting and socialising, eating and drinking, having fun, etc. Computer stations are to be distributed in clusters throughout the building and not grouped in one place any more.

More personalized services, such as individual consultation for term paper help, need more space near the traditional reference desk. When planning and arranging library classrooms with computers it is necessary to do it in a way that enables the interaction between the students and which avoids the instructor hiding himself behind a 'teaching bunker' of information technology.

But it is also necessary - as seen before in the case of LMS - to bring the library and librarians to people where they work and when they need information. 'Satellite sites' with space for individual consultation perhaps can also be arranged directly in departments or other commons space outside the library, so that the library as consultation agency is available campus-wide.

The libraries' educational role in respect of content: information literacy activities

1. Information literacy and its definition

Even at the beginning of the last century the Nobel laureate in chemistry - and one of the predecessors of many later efforts to improve the communication of scholarly information - Wilhelm Ostwald wrote: "It is not enough to found libraries. It is necessary, by means of lectures and bibliographic lists, to instruct those eager for knowledge in the best methods of utilizing their treasures. And this is by no means so easy as it sounds!" (Ostwald , 1911).[3]

In information management as well as in e-learning, there is a need for an increased awareness of intellectual property, since questions about copyright, patents, plagiarism etc. are frequently directed to library personnel. This is also a part of information literacy (IL), and is an important part of the library's educational role.

In business and industry there is not only a need for subject-specific knowledge but also for information literate people who can work in teams. Competencies in self-directed learning and informing, as well as in information selection and their use as outcomes of academic education, are absolutely essential for modern enterprises. IL is really "a prerequisite for the information society" as Sheila Webber stated in a talk at the Annual Meeting of the Deutsche Gesellschaft für Informationswissenschaft und Informationspraxis (DGI) 2005. According to Loyd IL is a meta- competence, which enables one to learn new skills and knowledge (Loyd, 2003). Webber and Johnston offer the following holistic definition: IL is "the adoption of appropriate information behaviour to identify, through whatever channel or medium, information well fitted to information needs, leading to wise and ethical use of information in society." (Webber & Johnston, 2003). Hans Roes described IL as the "core competence" of the "knowledge economy" (Roes, 2001).

There exist numerous definitions of IL and also discussions about using better terms like electronic literacy, information fluency, information skills etc. IL includes the ability to find and use information needed for work, study and research. In regard to a specific information system, it means the ability to use this system efficiently. The knowledge of information sources, the creativity to design an information process, as well as the ability to cope with information overload (selection of relevant information, structuring and retrieving information) is also often called information literacy. Taking Christine Bruce's view on IL as the "sum of the different ways it is experienced" (Bruce, 1999) the Swedish Ola Pilerot described IL education as "helping learners change/broaden their repertoire of experiences" (Pilerot, 2003, p.5).

From a constructivistic concept of learning, it is important not only to convey knowledge or abilities but to convey the different points of views of observers. Self observation and reflection on one's own learning process is necessary. IL activities of libraries should help one to become aware and to develop one's own style of how to deal with information. How do I actually inform myself? This paper proposes the following definition of IL: In addition to efficient retrieval and navigation strategies, information literacy is the creativity to organize and shape one's own information process in a conscious and demand-oriented way (Hapke, 2005). Didactic tasks are supporting, consulting, orientating, changing of attitudes, improving critical thinking as well as cooperating via the learning environment. But the teaching library - see the title of Lux and Sühl-Strohmenger who give a very good account of IL activities in German libraries (Lux & Sühl-Strohmenger, 2004) - may perhaps be not the right term for the future. Nobody wants to be taught, especially not by the library. Teaching today means learning empowering

Christine Pawley views IL also as "a matter of making enlightened and informed consumer choices" (Pawley, 2003). Today consumers of information are increasingly not passive but become active, they change between roles as reader and writer, consumer and producer of information, a fact that was already observed in 1934 by the German philosopher Walter Benjamin when he wrote: "For since writing gains in breadth what it loses in depth, the conventional distinction between author and public is disappearing in a socially desirable way... Literary competence is no longer founded in specialized training but is now based on polytechnical education, and thus becomes public property. It is, in a word, the literarization of the conditions of living that masters the otherwise insoluble antinomies..." (Benjamin, 2001)

SOME REMARKS ON ACTUAL ISSUES IN INFORMATION LITERACY

How will IL change with the changing electronic environment? For example: What are the implications of federated search tools for IL training, a question which was treated by a masters thesis in the University of Sheffield (McCaskie, 2004). An increased use of databases was observed but there were also concerns about the quality of results produced from these searches. So it is important to include federated search tools in IL training and tutorials to make patrons aware of differences between meta-search interfaces and the native search interface.

Studies about the outcome of IL activities are needed, e.g. studies which observe and prove that a person who has taken an information literacy course has developed into a more competent person. The problem here is how to measure this. One example for this is Limberg (Limberg, 1999).

It is also interesting to look at the topic of IL and its different specification in universities and business enterprises (Ingold, 2005b). Until now IL activities in universities have been too greatly library dominated and a view from outside can balance this. In the enterprise the search for information is less important than information use and production and less important than coping with information overload. With this view, you also come close to individual knowledge management issues. Often conscious delegation to the information professional is also part of IL. In turn to have this business view on IL can help the library to better promote and tailor their services to their clients.

It is useful also to reflect on critical voices on IL which give new insight and ideas. This was done by Ingold discussing the librarians' view of IL (Ingold, 2005a). In Germany, at least, we are too uncritical of our own activities. Is IL too much dominated by the library? Perhaps consultation is better than education as a library's role in IL. It is well not to forget to improve the information systems themselves before educating their users how to use them. What about IL and the growing importance of visual literacy (Marcum, 2004)? There are no easy answers or solutions in all of the issues mentioned above.

INFORMATION LITERACY AND LEARNING

For Mandy Lupton IL is a learning experience, and "the value of generic, standalone, parallel and foundation courses for IL education is dubious" (Lupton, 2004). IL is an intrinsic part of learning as a response to a context and not a characteristic of the learner. Experiencing IL is bound up in the topic, course and discipline. Real life only contains "micromoments" (Bruce, 2002) of searching information and of the use of information systems. There does not exist "the" context, "the" user or "the" system, but of real impact are "usings" (Dervin, 1996), the real use of a specific system through a specific user in a specific context. Just-in-time models are necessary for IL instruction. The first information source for a student is another student - at first they ask each other for help. This over-the-shoulder- learning is very important for a learner, especially for becoming familiar with a computer application. The possibility to do this over-the-shoulder-learning should be used also in electronic reference services: pushing web pages, surfing and exploring information sources jointly with the customer.

To address the "teachable moment" (Block, 2003) of our customers as well as the full complexity of IL, we need a wide and diverse range of activities to promote IL and reference: one-off sessions in-class or outside of class, online tutorials, just-in-time-support as virtual reference, face-to-face meetings, newsletters via email, bookmarks, leaflets etc. Our customers are very different as learning types. IL has to be embedded in the learning context and it is more than measurable abilities and knowledge which can be tested. Acquisition of information skills does not happen casually. Learning objects to develop information skills also have to be included in the full range of study skills and in the technical infrastructure like the virtual learning environment (Stubley, 2005).

IL ACTIVITIES OF THE TUHH LIBRARY

Engineering students normally use the library in their first two study years to consult the textbook collection or use it to study for examinations. One-off sessions integrated in lectures serve as a beginning to introduce students to library- based databases and services. For a number of years, in the TUHH there has been an agreement between the Academic Deanery in Chemical Engineering and the Subject Librarian for Chemical Engineering that the librarian gets the possibility for a one-off session in specific key lectures. With these activities it is possible to reach at least 80 to 90% of the process engineering students by the time they have graduated. In the 4P P term of study the lecture "Biochemical and biological foundations for engineers" begins with a module about research methods where library services are presented. A faculty member introduces the librarian by emphasizing the importance of information skills. This is the best that can happen for the library.

Nevertheless this presentation is not really in a good learning context because after it the students do not have to do any information research. However, the students acquire an important learning goal: in the library there is a person who can give subject specific consultation and help when searching for information. A better case is a course where the students have to prepare a presentation about designing a biotechnical process. In the "Process Design Course", offered every spring for three weeks, teams of students have the task of designing a whole chemical industrial process plant by considering materials, processes, security, environmental and economical issues. The librarian is part of the opening event and gives a presentation called "The world of engineering information - 10 points for survival". These points are:

1. Be conscious of your information behaviour.

2. Use tutorials, subject gateways and literature guides to inform yourself about how to search for information.

3. Use your local research library and consult a librarian or information specialist.

4. Use encyclopaedias and other reference works for preliminary orientation.

5. Play with search terms when exploring database features (Boolean logic, wildcard symbols, neighbourhood operators, search fields, ...)

6. Search for journal articles in subject-specific databases.

7. Don't forget to search for patents and data.

8. Evaluate your search results with respect to relevance as well as quality of the document you've found and think about processing your information.

9. Keep yourself up-to-date by browsing through journal contents, subscribing to mailing lists and reading weblogs.

10. Reflect on information ethics (intellectual property, copyright and plagiarism) and policy (ownership, privacy) as well as economics (commercial and open access).

A further result of this local faculty-librarian collaboration was the inclusion of an appendix with the title "The world of biotechnology information - 8 points for reflecting on your information behavior" in a biotechnology textbook (Buchholz, Kasche & Bornscheuer, 2005).[4]

Raising awareness is the most important issue in such events, but this is essentially not limited to students only, as the following example indicates. In a survey on the information behaviour of scientists or scholars done in Germany by Arthur D. Little a great uncertainty about the results when searching information was stated (Little, 2001). Here it is necessary to make clear that on the one hand uncertainty is part of every information process itself (Kuhlthau, 2004). When searching you always have to find a balance between finding too much or too little. On the other hand situations of uncertainty are common in electronic information systems, e.g. when you think of the quality of information on the web, its ownership, and the restriction of access (Kuhlen, 1999).

A further project of the TUHH library was to complement in-class activities with an online tutorial. With the bilingual online tutorial DISCUS (Developing Information Skills & Competence for University Students), which can be used independently of time and space, the TUHH library offers a playful and explorative way of transferring information skills (Bieler, Hapke & Marahrens, 2005). The user of digital libraries is viewed as an 'information player' (Nicholas & Dobrowolski, 2001) who plays with databases and search terms to improve research results. For learning, play is an important concept, which should be explored (Fister, 2005).

DISCUS also gives a subject-specific orientation and includes interactive and task-oriented elements, laying special emphasis on the visualization of the tutorial's interface. The high-quality visual presentation of the user interface creates an enjoyable ambience for learning and supports perception and orientation. It creates positive emotions and increases learning activities. Didactically, DISCUS presents the content in a manifold way, offering changes of perspectives, appealing to different learning types, and enabling the student to have fun. The (red) thread as leading metaphor is visible on every screen and also in the library.

An advertising bookmark for DISCUS says: "Grab the thread to information! The Online Tutorial DISCUS ...'cause Quicksearch is not enough!". The tutorial includes a "survival guide" for using databases, a knowledge base "DISCUS compact" as a systematically ordered, linear text (also downloadable as a PDF), and case studies for searching for information in biotechnology like citric acid, Keratin, alcohol, biological degradation and whey. In process engineering, the topic "Use of rapeseed oil in the car" gives the background for exercises to search for information.

Picture 3: The thread in the TUHH library

But which students do take the time to use the tutorial? This will only be done by those students who, in most cases, would have asked a librarian for help or who already know about the worth of IL for their theses or course work. To meet the challenge of too little usage of the tutorial the TUHH library is partner in a further project called BibTutor, which is funded by the German Federal Government. Together with three other university libraries (Darmstadt, Heidelberg, Kaiserslautern) and the German Research Center for

Artificial Intelligence in Kaiserslautern, we want to facilitate searching in the original search interface whether it is a library catalogue, a union catalogue or a subject- specific database.

In addition to support database selection, BibTutor will give context-specific advice when the user searches a specific database interface. It also should offer context-specific, just-in-time (e-)learning, through linking, for example, directly   into DISCUS. One important question of intelligent systems is which tasks of the user will be taken over by the system and what has the user to learn and to think of when using it. BibTutor should give the user the possibility to reflect on what s/he does when s/he wants to reflect. "Learning how to use databases is not only a technical but primarily a problem of social 'hermeneutics' (= the ability to ask critical questions, instead of just believing what is written or   programmed or stored)" (Capurro, 1990, p. 132).

In a further project called VISION (Virtual Services for Information Online), the TUHH library plans to create mini- tutorials for inclusion in LMS for topics like presenting information (citation styles, visualization), supporting electronic publication, and consulting in intellectual property and copyright issues. We also want to apply what we have learnt from DISCUS, the importance for visualization of information and interfaces. Furthermore smaller tutorials are perhaps more appropriate to user needs than such a comprehensive tutorial like DISCUS.

Organisational issues: the learning library

The most important reason for the TUHH library to do an e-learning project like the DISCUS IL tutorial was not that we thought that many students would take the time to use such a tutorial. Beneath having a landmark showing that the library does e-learning, the main reason for us was a strategic one: to become, as a library, part of the e-learning discussion within the university, to be visible as a library in the electronic learning environment. Enlarging cooperation   with faculty was another strategic goal.

Until now the TUHH - a small university which serves approx. 6000 students, 25% of whom are foreign students - has had no consistent vision or strategy for e-learning. What are found in most cases are unconnected activities in e- supported teaching. In Germany the universities have just begun to enter the area of e-learning, although there are many marketing activities under this headline. In Germany the introduction of bachelors and masters degrees combined with new curricula (including mediating key skills as explicit learning goals) offers opportunities for libraries to get involved in teaching information literacy and in e-learning.

This part of the paper discusses some of the organizational issues when taking the educational role of the university library seriously. Three aspects are important here: issues within the library, within the university, and between libraries in a library union.

Within its organisation the library has to be a learning organisation itself: collecting experiences from other libraries, data on users and their behaviour and ideas from the literature, relating this to its own institution and context as well as to its partners within its parent institution, transforming old and creating new services, new organisational structures and new relationships to old and new partners, and finally increasing its overall value to its patrons.

Educational services of university libraries also influence the organisational development of the library itself, e.g. through creating new special service departments and through cooperating with new partners, for example, the student services department, perhaps even merging the library with the computer centre or the educational centre, if one is present (Collier, 2005). Which way to choose depends on the local situation in the university, which is influenced not only by functional requirements but also by people and their political power. Especially in e-learning local structures determine the role taken by the library, and most approaches are locally based (Melling, 2005)

An interesting development is that of the learning commons as an extension of information commons and as a central facility that provides space, technology, and expertise to support learning. At the University of Guelph in Canada (Schmidt, 2005) it is a strategic alliance that includes the library to combine services in research, reference and IL, as well as computing, in teaching and instructional design, and in writing and learning support. The aim is to enhance learning, writing, research and technology skills. This seems to be an interesting way for the library to be an active partner in a collaborative learning process (Simon, 2004, p. 150). The British model of "educational informatics" may be an academic foundation for such a collaboration (Levy et al, 2004). Another way is the British "Learning Centre" model (Oysten, 2003). In Finland - see for example the Tritonia Learning Centre – there exist learning centres which are not only responsible for supporting student learning - and here especially with education in information skills - but also for supporting, consulting and training faculty when teaching. The e-learning expert Neil McLean from Instructional Management Systems Centre (IMS) in Australia speaks of different service domains in higher education: E-Research, Scholarly Information, E-Learning and Administrative Computing. What will be the role of the library in these domains in the future? (McLean, 2004)

The emergence of digital libraries forces library networks like the GBV Common Library Network in Northern Germany to develop strategies to cope with specific local demands for teaching and learning in the electronic world and to decide which local demands are also of interest for other network libraries and how to organise this cooperation. The GBV has developed a strategic working paper about the digital future of the library union. The challenge is to cope with the complexity and diversity of existing local and decentralised systems with varying technical solutions and context-specific resources, as well as to integrate them into central resource description systems, portals and other access points (GBV, 2005).

What is not yet really recognised by most libraries is the need for knowledge management instead of information management issues for digital library portals. The growing future of computer-supported cooperative work has to be integrated in library portals and services.

It is necessary to provide opportunities to make active and self-directed learning for students possible and to offer a place - real and virtual - where learners can meet in a community to work together and develop shared knowledge.

CONCLUSION: THE LEARNING FACILITATING LIBRARY

The examples above emphasise the consulting role of the library as the most important one, especially in education. The library has to be a learning library itself as an organisation to compete with the challenges of the future and to adapt itself to the manifold needs of its users and its parent organization. The teaching library is not the goal of library development as a service organization, but the enhancement of the learning empowering and facilitating library.

In the future university libraries as service institutions may become more and more part of research and learning. Consultation and media will be offered when needed at the point of use, in the laboratory, in the seminar or lecture or elsewhere. Hans Roes spoke of the rediscovery of the library in education. Universities have to combine their strategies for the development and improvement of learning and teaching with their concepts for the service departments (Roes, 2001).

A story about an exhibition in the Jewish Museum in Berlin last year concludes this paper. The presentation of the exhibition named "10+5 = God : the power of numbers and signs" was accompanied by the display of cards for a card game ('happy families', 'Quartett') in the exhibition rooms. The cards visualised and mobilised interesting information on the exhibition's topic. The visitor was subtly urged to collect and organise the cards to get a complete game and so on visiting all the parts of the exhibition. The cards were used as means to transfer knowledge and to educate the visitors. This example of play symbolises the way the university library can work to serve its customers in an educational role: offering its collections in an organised way and allowing patrons to mobilise its contents to educate themselves in a playful fashion.

ACKNOWLEDGEMENTS

Thanks for looking at the English to F. Bartow Culp, Head Librarian, Mellon Library of Chemistry, Purdue University,

W. Lafayette, IN

NOTES

1. According Daniel Boorstein cited by Haigh (2004, p.20)

2. Look at "Library as place: rethinking roles, rethinking space" (2005) of the American Council on Library and Information Resources (CLIR) for more stimulation.

3. For more on Ostwald and his activities as an information pioneer see Hapke (2003).

4. Appendix I, pp. 419-426. An enhanced online version can be found at http://www.tub.tu-harburg.de/2552.html?docinput[lang]=en

REFERENCES

1. Benjamin, Walter: "The newspaper". In: *Benjamin, Walter: Selected writings*. Cambridge, Mass. : Belknap Press, 2001, Vol. 2, pp. 741-742.

2. Bieler, Detlev, Thomas Hapke and Oliver Marahrens: "Lernen, Informationskompetenz und Visualisierung - das Online-Tutorial DISCUS (Developing Information Skills & Competence for University Students) der Universitätsbibliothek der TU Hamburg-Harburg". *ABI-Technik*, 25(2005)3.

3. Block, Marylaine: "Teach them while they're asking for information : reference as a teachable moment". In: *Net effects: how librarians can manage the unintended consequences of the Internet.* Medford, NJ : Information Today, Inc., 2003, pp. 76-79.

4. Bruce, Christine S.: "Experiences of Information literacy in the Workplace." *International Journal of Information Management,* 19(1999)1, 33-47.

5. Bruce, Harry: *The user's view of the Internet.* Lanham, Md. : Scarecrow Press, 2002.

6. Buchholz, Klaus; Volker Kasche and Uwe Th. Bornscheuer: Biocatalysts and enzyme technology. Weinheim : Wiley- VCH, 2005.

7. Capurro, Rafael: "Towards an information ecology". In: *Information quality : definitions and dimensions,* ed. by Irene Wormell. London : Taylor Graham, 1990, pp. 122-139.

8. Collier, Mel: "Convergence in Europe outside the United Kingdom". In: *Managing academic support services in universities : the convergence experience,* ed. by Terry Hanson. London : Facet, 2005, pp. 181-201.

9. Demas, Sam: "From the Ashes of Alexandria : What's Happening to the College Library?". In: *Library as Place: Rethinking Roles, Rethinking Space.* Washington, D.C. Council on Library and Information resources, 2005. http://www.clir.org/pubs /reports/pub129/demas.html

10. Dervin, Brenda: "Information Needs and Information Seeking: The Search For Questions Behind the Research Agenda". Paper at the Workshop "Social Aspects of Digital Libraries" in the University of California, Los Angeles, February 16-17, 1996, http://is.gseis.ucla.edu/research/dl/dervin.html

11. Dury, John: *The reformed Librarie-Keeper* (1650) / Introd. by Richard H. Popkin and Thomas F. Wright. Los Angeles: William Andrews Clark Memorial Library, 1983.

12. Fister, Barabara: "Smoke and Mirrors: Finding Order in a Chaotic World". Presented at the Workshop on Instruction in Library Use (WILU) 2005, University of Guelph, Ontario. http://homepages.gac.edu/~fister/WILU2005.html

13. Foucault, Michael: *Botschaften der Macht. Der Foucault-Reader: Diskurs und Medien* / ed. by Jan Engelmann. Stuttgart : Dt. Verl.-Anst., 1999.

14. "GBV digital - Das digitale Medien- und Serviceangebot im Gemeinsamen Bibliotheksverbund". *Mb : Mitteilungsblatt der Bibliotheken in Niedersachsen und Sachsen-Anhalt,* 130(2005)8-12. http://www.gbv.de/du/pdf/GBVdigital.pdf

15. Haigh, Thomas: "An Introduction to the History of Computing for the Computer Scientist." In: *Using history to teach computer science and related disciplines*, ed. Atsushi Akera and William Aspray. Washington, D.C.: Computing Research Association, 2004, pp. 5-26.

16. Hapke, Thomas: "'In-formation' - Informationskompetenz und Lernen im Zeitalter digitaler Bibliotheken." In: Bibliothekswissenschaft – quo vadis? = Library Science – quo vadis? : Eine Disziplin zwischen Traditionen und Visionen ; Programme – Modelle – Forschungsaufgaben, ed. Petra Hauke. München : Saur, 2005, pp. 115-130.

17. Hapke, Thomas: "From the world brain to the first transatlantic information dialogue : activities in information and documentation in Germany in the first part of the 20th century". *IFLA Journal*, 29(2003)4, 364-377. http://www.ifla.org/V/iflaj/ij-4-2003.pdf

18. Harley, Bruce; Dreger, Megan and Knobloch, Patricia: "The postmodern condition: students, the web, and academic library services". *Reference Services Review*, 29(2001)1, 23-32.

19. Ingold, Marianne: *Das bibliothekarische Konzept der Informationskompetenz : ein Überblick*. Berlin : Institut für Bibliothekswissenschaft der Humboldt-Universität zu Berlin,

2005a. http://www.ib.hu- berlin.de/~kumlau/handrei chungen/h128/

20. Ingold, Marianne: "Informationskompetenz: ein (neues) Leitbild für betriebliche Informationsstellen?" In: *Leitbild Informationskompetenz : Positionen, Praxis, Perspektiven im europäischen Wissensmarkt* ; 27. Online-Tagung der DGI, 57. Jahrestagung der DGI, Frankfurt am Main, 23. bis 25. Mai 2005 ; proceedings, ed. by Marlies Ockenfeld. Frankfurt am Main : DGI 2005b, pp. 15-26.

21. Klatt, Rüdiger et al.: Nutzung elektronischer wissenschaftlicher Information in der Hochschulausbildung". Eine Studie im Auftrag des Bundesministeriums für Bildung und Forschung, Projektträger Fachinformation / Sozialforschungsstelle Dortmund. Dortmund, 2001. http://www.stefi.de (Also available as: Elektronische Informationen in der Hochschulausbildung : innovative Mediennutzung im Lernalltag der Hochschulen / Rüdiger Klatt et al. Opladen : Leske + Budrich, 2001)

22. Kuhlen, Rainer: Die Konsequenzen von Informationsassistenten : was bedeutet informationelle Autonomie oder wie kann Vertrauen in elektronische Dienste in offenen Informationsmärkten gesichert werden? Frankfurt a.M. : Suhrkamp, 1999.

23. Kuhlthau, Carol Collier: *Seeking meaning : a process approach to library and information services.* 2nd ed. Westport, Conn. : Libraries Unlimited, 2004.

24. Levy, Philippa; Nigel Ford ...: "Educational informatics: an emerging research agenda". *Journal of Information Science,* 29(2003), 298-310

25. *Library as Place: Rethinking Roles, Rethinking Space.* Washington, D.C. : Council on Library and Information resources, 2005. http://www.clir.org/pubs/abstract/pub129abst.html

26. Limberg, Louise: "Experiencing information seeking and learning: a study of the interaction between two phenomena". *Information Research*, 5(1999)1. http://informationr.net/ir/5-1/paper68.html

27. Arthur D. Little: *Zukunft der wissenschaftlichen und technischen Information in Deutschland : Ergebnisse der empirischen Untersuchungen über das Informationsverhalten von Wissenschaftlern und Unternehmen.*

28. Zwischenbericht an das Bundesministerium für Bildung und Forschung. Oktober 2001. http://www.bmbf.de/pub/zukunft_der_wti_in_deutschland.pdf

29. Loyd, Annemaree: "Information literacy : the meta-competency of the knowledge economy? An exploratory paper". *Journal of Librarianship and Information Science*, 35 (2003)2, 87-92.

30. Lupton, Mandy: The learning connection: information literacy and the student experience. Adelaide : Auslib Pr., 2004.

31. Lux, Claudia and Wilfried Sühl-Strohmenger: Teaching library in Deutschland : Vermittlung von Informations- und Medienkompetenz als Kernaufgabe für Öffentliche und Wissenschaftliche Bibliotheken. Wiesbaden : Dinges & Frick, 2004.

32. Lynch, Clifford: "The Afterlives of Courses on the Network : Information Management Issues for Learning Management Systems". *EDUCAUSE Research Bulletin*, (2002)23, 14 p. http://www.cni.org/staff/cliffpubs/ECARpaper2002.pdf

33. Marcum, James W.: "Beyond Visual Culture: The Challenge of Visual Ecology". *portal:Libraries and the Academy*, 2(2002)2, 189-206.

34. McCaskie, Lucy: What are the implications for information literacy training for higher education with the introduction of federated search tools? MA (Librarianship) dissertation. University of Sheffield, Department of Information Studies, 2004. http://dagda.shef.ac.uk/dissertations/2003-

04/External/McCaskie_Lucy_MALib.pdf

35.   McLean, Neil: "The Ecology of Repository Services: A Cosmic View". 8P   P   European Conference on Research and Advanced Technology for Digital Libraries (ECDL 2004), University of Bath, UK. http://www.ecdl2004.org /presentations/mclean/

36.   McLean, Neil and Clifford Lynch: Interoperability between Library Information Services and Learning Environments Bridging the Gaps : A Joint White Paper on behalf of the IMS Global Learning Consortium and the Coalition for Networked Information. May 10, 2004. http://www.imsglobal.org/ digitalrepositories/CNIandIMS_2004.pdf

37.   McLean, Neil and Heike Sander: *Libraries and the Enhancement of E-learning* (OCLC White Paper). Dublin, Ohio: OCLC, 2003. http://www.oclc.org/index/elearning/default.htm

38.   Melling, Maxine (Ed.). Supporting e-learning: a guide for library and information managers. London : Facet, 2005

39.   Nicholas, David and Tom Dobrowolski: "The 'information player' : a new and timely term for the digital information   user". In: *Handbook of information management*, ed. by Alison Scammell. 8. ed. London : Association for Information Management. London : Aslib-IMI, 2001, pp. 513-522.

40.   Oblinger, Diane G. and James L. Oblinger (Eds.). *Educating the net generation*. Boulder, Co.: EDUCAUSE, 2005. http://www.educause.edu/educatingthenetgen

41.   O'Day, Vicki L. and Bonnie A. Nardi: "An ecological perspective on digital libraries" In: *Digital library use : social practice in design and evaluation*, ed. Ann Peterson Bishop. Cambridge, Mass. : MIT Press, 2003, pp. 65-82.

42.   Ostwald, Wilhelm: "Biology of the savant : a study in the psychology of personality". *Scientific American*, Supplement No. 1862, 9 September 1911, 169-171

43. Oyston, Edward (Ed.). Centred on learning : academic case studies on learning centre development. Aldershot : Ashgate, 2003.

44. Pawley, Christine: "Information literacy: a contradictory coupling." *The Library Quarterly* 73(2003) 422-452.

45. Pilerot, Ola: "Information literacy at a distance - collaboration between a university library and two public libraries". In: *Second International Conference on Information and IT Literacy*. Glasgow Caledonian University, 2003. http://www.elit-conf.org/elit2003/papers/ppt/pilerot.pdf

46. Poole, William: The university library and the university curriculum. Chicago : Revell, 1894.

47. Roes, Hans: "Digital Libraries and Education : Trends and Opportunities". *D-Lib Magazine*, 7 (2001) 7/8, http://www.dlib.org/dlib/july01/roes/07roes.html

48. Schmidt, Nancy: "Putting the 'learning' in learning commons : a student services perspective". In: *Building the Future: Designing Academic Libraries as Learning Spaces*. CIC Center for Library Initiatives Conference, May 2-3, 2005, University of Chicago. http://www.cic.uiuc.edu/programs/CenterForLibraryInitiatives/Archive/ConferencePresentation/LibrarySpacesConference2005/LearningCommonsPresentationNancySchmidt.ppt

49. Shneiderman, Ben: Leonardo's Laptop: human needs and the new computing technologies. Cambridge, Mass. : MIT Press, 2002.

50. Simon, Theresia Maria: Die Positionierung einer Universitäts- und Hochschulbibliothek in der Wissensgesellschaft : eine bibliothekspolitische und strategische Betrachtung. University of Potsdam, Dissertation, 2004.

51. Stubley, Peter: "Just one piece in the jigsaw: e-literacy". In: *Supporting e-learning: a guide for library and information managers*, Melling, Maxine (Ed.). London : Facet, 2005, pp. 113-137.

52. Tully, Claus J.: "Nutzung jenseits systematischer Aneignung - Informalisierung und Kontextualisierung". In: *Verändertes Lernen in modernen technisierten Welten: organisierter und informeller Kompetenzerwerb Jugendlicher.* Wiesbaden : VS Verlag für Sozialwissenschaften, 2004, pp. 27-55.

53. Twidale, Michael B. and Karen Ruhleder: "Over-the-shoulder-learning in a distance education environment". In: *Learning, culture, and community in online education : research and practice,* Caroline Haythornthwaite & Michelle

54. M. Kazmer, eds. New York, NY : Lang, 2004, pp. 177-194.

55. Virkus, Silje and Silvi Metsar: "General introduction to the role of the library for university education". *LIBER Quarterly,* 14(2004)3/4, 290-305. http://liber.library.uu.nl/cgi-bin/pw.cgi/articles/000101/index.html

56. Webber, Sheila and Bill Johnston: "Assessment for information literacy: vision and reality. In: *Information literacy and IT literacy : enabling learning in the 21st century,* ed. by Allan Martin and Hannelore Rader. London : Facet 2003, pp. 101-11

Chapter 5

Data standardization in Digital Libraries: An ETD Case in Turkey

Özlem Şenyurt Topçu*, Tolga Çakmak, Güleda Doğan

Hacettepe University, Faculty of Letters, Department of Information Management, Ankara, 06800, Turkey

ABSTRACT

Nowadays, data integrity and data standardization are significant topics for information retrieval systems and also for digital libraries. Although, many standards (such as VIAF, AACR2 and MARC) and institutional regulations developed for data standardization by the nature of library and information science field, consistency between different content resources is still a problem for today's information systems. It is also one of the most important steps for digital library projects especially for the ones who are planning to gather data from different content resources. As one of these projects, a digital library for electronic thesis and dissertations (ETD) for LIS in Turkey has been developed via a digital library project executed by PhD students of Hacettepe University Department of Information Management. In this direction, this paper aims to explain data standardization processes and limitations due to regulations, language and cultural characteristics of the country with examples from the indicated project. As a result of the project a standardized structure for ETD systems were created. In this context, author and advisor names, Turkish and English titles, keywords, access restrictions were determined as the main elements for standardization processes. In the end of the project, an authority file was created for advisors via VIAF, RDA and AACR2 in order to improve efficiency of access points by ensuring data integrity

Keywords: Data integrity; data standardization; ETDs; digital libraries; Turkey

INTRODUCTION

Of late, the initiatives about digital libraries and digital archives have increased especially in universities and academic institutions. Furthermore, it is also seen that many digital archives have been developed with the aims of preserving cultural heritage and providing access to cultural heritage assets via recent technologies. Bibliographic description of information resources and cultural heritage assets covered by digital libraries and archives is an essential process in order to meet information needs of users. It also provides efficiency for information retrieval and makes digital library and archives usable for users by supporting decision-making processes. On the other hand, standardized presentation of data identified in digital library and archives is as important as description of metadata fields. It is an important requirement for the initiatives especially whose aim is to create a single platform for different content providers. Based on this information, this study addresses the data standardization phases in the case of a local digital archive project whose main objective is to develop an ETD platform specialized in Library and Information Science in Turkey by gathering data from different content providers.

DATA STANDARDIZATION, CONSISTENCY AND CLEANING

Data standardization provides consistency between the content and format of data types represented in a system. Furthermore, data standardization facilitates efficient consistency for data mapping and data outputs (IBM, 2013). According to the literature data standardization, consistency and cleaning topics are covered by different fields like computer science, statistics and information science. In this context, different terms like normalization, data cleaning, noisy data are used to describe data standardization studies. It is indicated that data standardization processes improve information retrieval, prevent loss of data and provide unduplicated data (Normalleştirme, 2013; Tyagi and Samuel, 2013).

Data standardization is a determinative fact about data quality and high quality data should have some criteria. These criteria are accuracy, integrity, completeness, validity, consistency, uniformity and density data quality (Tyagi and Samuel, 2013). Data standardization provides internal consistency of data represented in digital libraries and archives. Through this process, each data type described in the digital library and archive have the same content representation properties (IBM InfoSphere Information Server, 2013). Standardization processes of data are referred as one of the essential processes in terms of user interaction and meeting information needs. Furthermore data standardization processes increase the clarity level of data to support users' interaction with digital library and archive systems in terms of accessing to required information (Tyagi and Samuel, 2013).

The necessity of synchronized structures of data presented in digital library and archives is discussed in studies. Accordingly many studies are conducted to provide data consistency during the implementation phases of digital library and archives. In this context data consistency is defined as a concept that summarizes validity, usability, integrity of applications and data in digital library or archive systems. Digital library or archive systems that have consistent data present a high quality and consistent platform for their users (Data Consistency, 2013).

Conceptually, data standardization is relevant with many components. One of these components is data cleaning processes. Data cleaning can be described as a set of operations carried out for detecting and removing errors and inconsistencies from data (Rahm and Do, 2000). Main aim of data cleaning is to weed out unsuitable or incorrectly entered data in the data set (Tekerek, 2011). Data cleaning is also stated as a process that increases data quality and solves data quality problems. According to Ramd and Do (2000, p.3), data quality problems consist of single-source problems and multi-source problems.

In their study, Ramd and Do (2000, p.3) divide data quality problems into two levels. These levels are metadata schema level and instance level. On the other hand, misspellings, redundancy of data, contradictory values are the essential problems that require data standardization and data cleaning operations. It is also expressed that data cleaning processes include completion of missing data, ensuring data consistency via identification of outliers (Oğuzlar, 2003).

DATA STANDARDIZATION AND METADATA

The need for standardization of data practices has step up with the growing of digitization technology in library related operations. The creation of digital collections using digital library technologies, either in the form of 'born- digital' or migrated into digital form, is now an important part of the activities for most of the higher education institutions. Using of these digital collections effectively is dependent on the metadata quality. Furthermore, management of digital resources requires standardized and high quality data (Gartner, 2008, p. 4).

After describing metadata fields for a digital archive/library, before the data entrance processes, data need to be standardized. As a core element of information retrieval systems metadata describes data quality and determines maintenance and preservation of a digital library. Data standardization is crucial for ensuring consistency with metadata for such reasons. Standardized data utilizes efficient discovery, access, transfer and use of common terms, common definitions, etc. (Gartner, 2008, p. 5; Why, 2013). Accordingly, standardized metadata enables users to access data effectively and efficiently by using a set of terminology (GeoNetwork, 2008, p. 32). Standardization of data and so metadata provide users finding data they need effectively and efficiently (Xie and Shibasaki, 2013).

With the advancements in technology, there are many digital library and archive systems implemented by the academic institutions. It is seen that many digital library and archives have the same metadata fields however they have different content data. In this direction, ISO/IEC 11179 (Specification and standardization of data elements) standard is stated as the most important outstanding standard for data representation in information systems by the aim of providing understandable and shareable representation of data stored in information systems. Moreover there are many rules and regulations developed by the libraries and library related communities and associations. These rules can be listed as: Cutter's rules, ALA Rules in 1908 and 1941, AACR (1967), AACR2 (1978), ISBD, AACR2R and RDA. Beside these developments, with the web archiving approaches, many metadata element sets were developed for data representation. In this context it is stated that Dublin Core based systems provide many advantages for data representation and interoperability of the systems (Caplan, 2003, p. 40, 55, 85). Additionally, there are many national and international data standardization projects especially for authority files for library automation systems. In this context, an international authority file project titled Virtual Authority File (VIAF) was created by OCLC (VIAF, 2013). As another project titled ORCID, it is aimed to provide a persistent digital identifier that distinguishes researchers from other researchers with a unique id (ORCID, 2013). This project can also be regarded as a data standardization effort for the identification of researchers and authors.

THE CASE OF LIBRARY AND INFORMATION SCIENCE ELECTRONIC THESIS AND DISSERTATIONS ARCHIVE IN TURKEY

In this part of the study, data standardization processes of a digital archive project titled Turkey Library and Information Science Departments Thesis and Dissertations Archive are described. There are 12 Library and Information Science (LIS) departments in Turkey. Only four of them have master and PhD education. In this respect, thesis and dissertations that are used for the project were gathered from four universities. These universities are Ankara University, Istanbul University, Hacettepe University and Marmara University (BBY Haber, 2013; Düzyol, 2011, pp. 4-5).

The main aim of the project is to provide a platform that contains all thesis and dissertations completed in LIS departments in Turkey. In parallel with this aim, objectives of the project are:

- Creating a union catalog for ETDs completed in LIS departments in Turkey.

- Developing a digital archive that presents full texts and bibliographic descriptions of all ETDs in LIS departments in a single platform

- Identification of ETDs via interoperable and standardized structures.

- Increasing access and visibility of ETDs via a digital library platform that supports OAI-PMH standards and protocols, and provides an interoperable environment for search engines and crawlers of similar digital   archives.

Content & Data Structure

Electronic collection presented in the digital archive consists of 436 post-graduate (masters and doctorate)   theses that are completed in LIS Departments in Turkey by the end of 2012. Table 1 shows the contribution of the   universities that are the content providers in the project.

Table 1. Content distribution of ETD platform.

Universities	MA		PhD		Total	
	N	%	n	%	n	%
Ankara University	62	18	29	30	91	21
Hacettepe University	126	37	36	37	162	37
İstanbul University	91	27	21	22	112	26
Marmara University	60	18	11	11	71	16
Total	339	100	97	100	436	100

As it is seen in Table 1, more than one-third (37%) of the collection presented in the digital archive is provided   by Hacettepe University. It is followed by İstanbul University (26%) and Ankara University (21%). Moreover, It is also remarkable contribution that 30% of the PhD theses provided by Ankara University. Plus, two-third of all   PhD theses is provided by Hacettepe and Ankara universities. As a consequence, data standardization processes were applied to 436 ETDs provided from four different content providers.

Data Standardization processes and data standardization work-flow

Data standardization is required to integrated presentation of ETD's that are situated in different sources with different structures/systems. Data presented under the metadata fields are standardized through various stages. Figure 1 displays the main data standardization workflow for digital library and archive initiatives. In this project, standardization workflow steps presented in Figure 1 were followed.

There are four Library and Information Science departments in Turkey that provide Master and PhD education. Therefore, these four departments are content providers for the data presented in the digital archive and the archive contains all theses and dissertations completed in these four departments by the end of 2012. Theses and dissertations that form digital archive are in PDF format. Finereader OCR (Optical Character Recognition) program were used just for the theses that cannot be copied because of the image based PDF files. In this framework, firstly data set was determined and created via various supplementary resources. These resources are a master thesis completed in 2011 (Düzyol, 2011), National Theses Center of The Council of Higher Education, institutional repositories and library catalogs and databases.

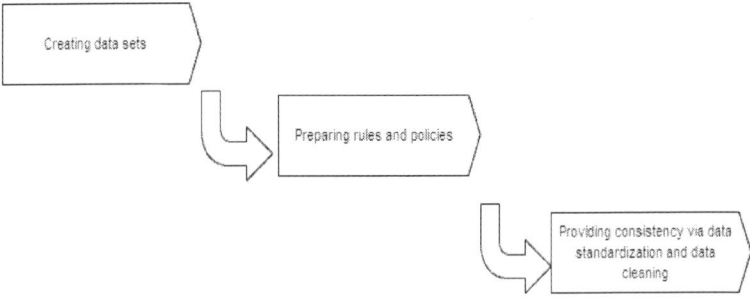

Figure 1. Standardization workflow

In the second stage of the project, rules and norms that will be used for the standardization processes were   determined. Policies were also created according to metadata fields and information resource characteristics of   thesis and dissertations. Some metadata fields were also qualified with effect of language and other cultural   characteristics of Turkey. AACR2 rules, VIAF structures and "ETD-MS v1.1: an Interoperability Metadata Standard for Electronic Theses and Dissertations" published by Networked Digital Library of Thesis and   Dissertation (NDLTD) are the main indicators for data standardization and presentation processes in the project.   Best practices and related national and international projects were also reviewed in this stage. In the light of these   resources and indicators, qualified metadata fields can be listed as ID, Name, Surname, Type, Year, Proper Title,   Alternative Title, Advisor, Link for full-text, Summary in Turkish, Summary, Access Restriction of National Theses Center of The Council of Higher Education, Keywords in Turkish, Keywords. In the end of the workflow,   data integrity were provided, and false and flawed data were corrected. Data standardization processes that were   carried out according to metadata fields are listed and summarized below:

ID: Every thesis and dissertations has an ID number that makes them unique in the digital archive. ID numbers were used to match bibliographic records of resources with their PDF based objects stored by the digital archive.   Besides the ID numbers, handle server system was also created via handle.net and every record defined by a   prefix nominated to Hacettepe University who is the official owner of the project. In this context, Registry of Open Access Repositories (ROAR) system registry was also completed.

Authorities: Virtual Authority File (VIAF) were used for author names as a directive resource. Besides, records were displayed in the form of last name, name within the framework of author entries specified by   AACR2. Longest versions of the names were used for author and advisor names (i.e. Yaşar Ahmet Tonta instead of Yaşar Tonta or Yaşar A. Tonta). Current names preferred for the women authors and advisors that are married/divorced (i.e. Güleda Doğan instead of Güleda Düzyol). Only first letters of the titles are capital as it is in AACR2 except proper names in the titles.

Title: Titles and alternative titles of the thesis and dissertations were written in the form of only first letter capital unless there is a proper name as described in AACR2.

Date: Publication year information was identified in the form of month-day-year. The same format was applied for the thesis that has embargo date.

Keywords and subject headings: Capital letters were only used for the first letters of keywords unless there is a proper name as in titles.

Summaries: OCR and typo sourced errors in summaries were reviewed and corrected according data identification forms provided by the digital archive.

Usage restrictions: Usage restriction information of theses described by the Council of Higher Education was defined as authorized, unauthorized, restricted access options. Restricted access expression was used for the theses and dissertations that are not allowed to access for 1-2 or 3 years.

CONCLUSION

Digital libraries and archives are important structures for access to information resources and maintenance of the cultural heritage assets. These systems provide many opportunities for their owners and users. These opportunities are compliance with standards, interoperability with other systems, mostly supporting open access and scholarly communication. As essential components of these systems, Metadata fields and data represented in these fields provide effective information retrieval and support critical thinking processes of users. They are more important for the platforms that contain data from different resources and repositories. In this context, policies, rules/norms and studies based on providing data consistency have a vital role for digital library and archives in order to improve the effectiveness of information retrieval.

Although, there are five active departments, Turkey has a great potential with its new LIS departments for new thesis and dissertations. Additionally it would not be wrong to say that ETD initiatives in LIS field will potentially have a larger number of resources in the country. As the first union ETD initiative in LIS science, Turkey Library and Information Science Thesis and Dissertations Archive with its standardized and high quality data structures is a model for such attempts. In this regard this study reflects the importance of the efforts about data cleaning, data integrity and validity as well as the importance of compliance with standards, policies, rules and norms.

REFERENCES

1. BBY Haber (2013). *BBY Bölümler*. Retrieved March 29, 2013, Available at http://www.bbyhaber.com/bby/bby-bolumler/ Caplan, P. (2003). *Metadata fundamentals for all librarians*. Chicago: American Library Association

2. Data Consistency. (2013). Retrieved July 16, 2013, Available at http://en.wikipedia.org/wiki/Data_consistency

3. Düzyol, G. (2011). *Türkiye Kütüphanecilik ve Bilgibilim literatürünün entellektüel haritasının çıkarılması: bir yazar ortak atıf analizi çalışması*. Unpublished Master Thesis. Hacettepe Üniversitesi, Ankara.

4. Gartner, R. (2008). *Metadata for digital libraries: state of the art and future directions*. JISC Technology & Standarts Watch. Retrieved June 17, 2013, Available at http://www.jisc.ac.uk/media/documents/techwatch/tsw_0801 pdf.pdf

5. GeoNetwork Opensource (2008). *The complete manual.* Retrieved June 23, 2013, from http://apps.who.int /geonetwork/docs/Manual.pdf IBM (2013). *Making data consistent through standardization.* Retrieved July 24, 2013, Available at

6. http://pic.dhe.ibm.com/infocenter/iisinfsv/v8r5/index.jsp?topi c=%2Fcom.ibm.swg.im.iis.qs.ug.doc%2Ftopics%2Fc_Conformi ng_output_ data.html

7. IBM InfoSphere Information Server (2013). *Standardizing data.* Retrieved July 20, 2013, Available at http://pic.dhe.ibm.com/infocenter/iisinfsv/v8r5/index.jsp?topi c=%2Fcom.ibm.swg.im.iis.qs.ug.doc%2Ftopics%2Fc_Conformi ng_output_ data.html

8. Normalleştirme. (2013). Retrieved July 10, 2013, Available at http://yunus.hacettepe.edu.tr/~uras02/Hacettepe/3.sinif/Bilgis ayar/access/MIS_Dersnotu.pdf

9. Oğuzlar, A. (2003). Veri ön işleme. *Erciyes Üniversitesi İktisadi ve İdari Bilimler Fakültesi Dergisi, 21*(Temmuz-Aralık), 67-76. ORCID (2013). *ORCID.* Retrieved June 29, 2013, from http://orcid.org/

10. Rahm, E. and Do. H. H. (2000). *Data Cleaning: problems and current approaches.* Retrieved June 10, 2013, Available at http://dc-   pubs.dbs.uni-leipzig.de/files/Rahm2000DataCleaningProblemsand.pdf

11. Tekerek, A. (2011). Veri madenciliği süreçleri ve açık kaynak kodlu veri madenciliği araçları. paper presented at *Akademik Bilişim 2011 Konferansı.* Malatya: İnönü University.

12. Tyagi, B.K. and Samuel, P.P. (2013). *Data consistency, completeness and cleaning.* Retrieved June 25, 2013, Available at http://www.inclentrust.org/resources/PPS%202-DATA_ CONSISTENCY_Integrated.pdf

13. VIAF (2013). *Virtual Authority File.* Retrieved June 29, 2013, Available at http://www.oclc.org/viaf.en.html

14. Xie, R. and Shibasaki, R. (2013).Standardization framework for CEOP metadata development and application. *CEOP/IGWCO Joint Meeting.*

15. University of Tokyo, Japan. Retrieved June 20, 2013, Available at http://jaxa.ceos.org/wtf_ceop/documents/CEOP_Metadata _Report_20th.pdf

16. Why standardize metadata? (2013). Retrieved June 21, 2013, Available at http://gep.frec.vt.edu/pdfFiles/Metadata_PDF's/ 3.0MD_Presentation-Section3.pdf

Chapter 6

Environmental Friendly School Libraries as Excellence Resource Center in Creating Human Capital and Learned Malaysia Young Generation

Zailani Shafie[*], Nurul Huda Md Yatim, Razifah Othman

Faculty of Information Management, Universiti Teknologi MARA, Johor, Malaysia

ABSTRACT

Students can benefit from an ergonomically designed library and meaningful library programs as evidenced by a descriptive research study conducted at the University of Teknologi MARA, Johor Branch. Based on the data provided by Information Management students, significant differences were found in the following dimensions: (a) availability of efficient staff members, (b) comfortable studying atmosphere, (c) access to website for locating resources, and (d) needed professionally trained, full-time librarians. On the basis of these findings, it is recommended that user-friendly mobile digitalized school library be developed and qualified, full-time librarians be employed to manage it.

Keywords: School library; school research center; facilities; resources

INTRODUCTION

According to the National Institute of Building Sciences (2012), school libraries differ from most other types of libraries because they are contained within school buildings, which, in addition to library space, may include classrooms, auditoriums, circulation space, administrative offices, and cafeterias. In Malaysia, school library or normally known as School Resource Center (SRC) is a part of the whole continuum of educational provisions. The SRC, which forms an integral part in any school today, has

come a long way. It began as early as the first school, which was built by the British in the early 19th century. It was then called "khutub khanah" where books were stored in a corner of the school. When the small corner or room expanded into a bigger room and housed a bigger collection, the name of "khutub khanah" seemed outdated, so it was named "perpustakaan" (library). The library merely consisted of a collection of books to meet the students' reading needs as well as educational and language development. Hence, almost all schools in Malaysia have their own libraries. However, their very existence depends upon many factors such as availability of space, financial support or grant which in turn depends upon the school enrollment, and staff (Jusoh, 2002).

Explosive growths of technology bring about access of information beyond imagination. Many countries are moving along and embracing current technology in providing library services such as Cloud Storage or WorldCat, the world's largest online database for discovery of library resources. Most of them are tirelessly improving with a purpose of providing world's information and at a reduced cost by sharing hardware, services, and data, as opposed to traditional library management systems that offer individual libraries hosted, but siloed, hardware, software, and data storage. Malaysia has a long way to go in the field of library services and research particularly in school settings. Well-funded libraries have shown to bring positive impact on students' learning. One country in particular, Canada, has conducted much research, especially in the effects of well-funded libraries and student achievements. One notably study conducted by Lance (2004) and published by the Ontario library Association (2006), compared the analyses for state evaluations in Colorado, Alaska, Iowa, Pennsylvania, Texas, Oregon, Massachusetts, and New Mexico schools. The findings indicated students in schools with well-staffed, -stocked and - funded libraries scored from 10% to 25% higher on standardized tests than students in schools with poorly resourced libraries. A study of Massachusetts schools in 2000 concluded that students in schools with full-time teacher-librarians achieved higher on standardized tests than students in schools without full-time teacher-librarians.

Over 50 years after Malaysia gained its independence, rural school libraries in the country still in the old system. Furthermore, the new schools constructed lately also follow the same traditional model of school libraries. Therefore, there is an urgent need in restructure a new model library namely, mobile digitalized libraries, to cater the 21st century students and citizens beyond. This study seeks to identify the environmental friendly school libraries as excellence school resource center in creating human capital and learned Malaysia young generation.

LITERATURE REVIEW

In order to encourage people to come to the school library, it must be equipped with good facilities. Facilities can be defined as something designed, built, installed, etc., to serve a specific function affording a convenience or service (Facility, 2011). According to study in Illinois high schools, 11th grade ACT scores are highest when there is a high degree of true collaboration between library media specialists and classroom teachers in a wide spectrum of activities. It has been concluded that for schools to benefit as much as possible from strong libraries, access to them needs to be as flexible as possible, enabling teachers and students to work with the library media specialist and other staff and use the library as a classroom or study space as needed. (Lance, Rodney, & Hamilton-Pennel, 2005).

In any organizations, human resource plays as a vital role and an asset to act as a successor of an organization. Therefore, in order to be a successful school library, there is a need to have excellent and qualified staff especially those who have experience. When discussing about staff, study by Raja Abdullah and Saidina Omar (2003) found that almost all school libraries are managed in an ad-hoc manner while the teacher librarian position has been regularly changed and the time allocated has been at the average of about six hours per week only. With that limited time allocated, it is assumed that there is nothing much the teacher librarian can do in particular to meet the minimum standard of the school and library resource center. In Malaysia, the position of a professional teacher-librarian still remains unfilled, and their training and experience is far different from those acquired through the mainstream information professional. Otherwise, in order to attract those to fulfilled this position, it is also important to be seen from the incentives given. A study by Kamal and Othman (2012) suggests that another way to improve the effectiveness of school resource center is to ensure all schools in Malaysia employ full-time and adequately trained professional school librarians. According to Singh (2008), the Human Resource Practices enabler addresses the issues related to human resource management such as incentive schemes to attract and retain high caliber staff, as well as training programs to upgrade the skills and develop relevant competencies of new and existing staff.

External supports are essential to creating a successful school resource center. A study carried out by Fatimah (2002), has shown some school libraries were supported by one clerical staff and library assistants sponsored by the Parents Teachers' Association. To rectify this problem, the 2001 "Hala Tuju Pusat Sumber Sekolah", a proposal was made for full-time school libraries coordinators and full-time trained library assistants. Training of teacher-librarians is carried out by the Teachers Training Division, the Educational Technology Division and its network of 14 State Educational Resource Centers and the 367 Teachers Activity Centers. All the 27 teacher training colleges that offer pre-service programs have included the resource management component in their programs. The Diploma Program (3 Years), for example, have made it compulsory for students to take courses like Resource Management which include ICT Educational Technology and Library Science which is 105 contact hours. The Certificate Program (one year) offers the subject Information Technology Skills with only 45 contact hours. With these basic information and knowledge, the teachers are not only able to manage the school libraries or SRC, but more importantly, they can translate the resource-based learning concept into the classroom. There are pros and cons to this model.

The availability of professionally-trained librarians would allow teachers to focus on teaching and managerial responsibilities. Professionally-trained librarians take care of checking books in and out, processing new and selected books, managing magazines, taking care of requests to and from other schools, adding classes to the library calendar, assisting students visiting the library with basic help, and monitoring the library when teachers have to be in the classrooms or creating their lessons or even supporting teachers in their lesson planning. Major changes such as weeding, genre categorizing in fiction, and subject categorizing in nonfiction to library collection are possible, too. Professionally-trained and full-time librarians are able to create and sustain programs that promote reading, like book clubs, reading rewards programs, and reading events in classrooms. As a result, support staff is essential to creating a meaningful library program.

The growth of web technology has evolved in a way that it changes and empowers students and teachers alike to powerful tools and inquiry capabilities. Web-based technology, such as Web 2.0, with its dynamic, participatory, interactive, and social elements, helps transform web-based information into meaningful, motivating learning experience for users. This new way of information seeking behavior must be embraced if schools want to stay abreast with the rest of learning societies. Availability of support, access, and Information Technology (IT) tools with access to online services and resources will enable school libraries to address all information technology issues and needs. The government, Ministry of Education, and administrators need to ensure that the IT related facilities and software are made available to all schools, rural and urban, to search, store, disseminate and utilize information such that the conversion to knowledge is facilitated and made a reality.

This enabler ensures that there is adequate network infrastructure to ensure connectivity within the schools and with the outside world. As a result, appropriate information can be acquired efficiently and effectively from the right sources (Singh, 2005).

RESEARCH OBJECTIVES

The objectives of this study were to facilitate a thorough and comprehensive understanding of school libraries especially the rural areas in Malaysia; and to understand the challenges of those libraries in providing meaningful and effective services to students and future generation. Specifically, the study focused on the following research questions, to identify the:

- Facilities and designs in existing school libraries

- Type of technology, staff experience, and support available for
- students Conducive and pleasant environment needed by young
- generation Relationship between good environment and students'
- performance Effect of environment and sustainable design in students' performance

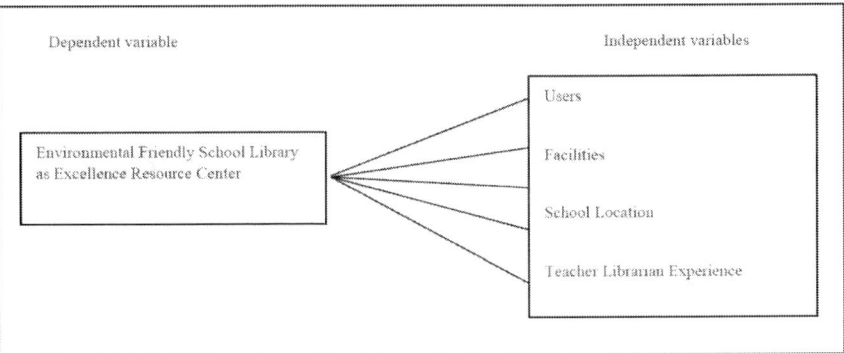

Figure 1. Theoretical framework

METHODOLOGY

The researchers created a survey to administer to Information Management (IM) students, which was designed to identify their perceptions of their experiences and challenges in using school libraries. The survey consisted of a 5-point Likert-type scale ranging from strongly agree to strongly disagree in four dimensions of school library: (a) physical facilities and designs, (b) managerial strategy and quality, (c) reading and learning culture, and (d) resources. The data were collected by randomly selected 80 students from the population of IM students at Universiti Teknologi MARA, Johor Branch. The study selected those students based on the following characteristics:

- Just finished their secondary school
- Still current with the environment of their school libraries
- Information Management Students who learned about School Libraries and Academic Libraries Therefore, selected subjects met the criteria used in the theoretical framework. Specifically, the subjects represented students from rural, urban, and boarding schools in their educational preparation.

The surveys were administered in a classroom environment. A total of 80 surveys were returned for a response rate of 100%.

RESULTS

According to Hair, Black, Babin, Anderson, & Tatham (2006), the closer Cronbach's Alpha is to 1.00, the higher is the reliability of the measures. The generally agreed upon lower limit for Cronbach's Alpha is .70 although it may decrease to .60 in exploratory research. The Cronbach's Alpha value is high found to be 0.833; therefore the reliability of the questionnaire is acceptable.

Descriptive statistics

Table 1. Summary of respondents' characteristics

1.1 Gender		Frequency	Percent	Valid Percent	Cumulative Percent
Valid	Male	24	29.6	30.0	30.0
	Female	56	69.1	70.0	100.0
Total		80	100.0		
1.2 School area		Frequency	Percent	Valid Percent	Cumulative Percent
Valid	School (urban area)	65	81.25	81.3	81.3
	School (rural area)	13	16.2	16.2	97.5
	Boarding School	2	2.5	2.5	100.0
Total		80	100.0		
1.3 Frequency to School Library		Frequency	Percent	Valid Percent	Cumulative Percent
Valid	>5 times a week	2	2.5	2.5	2.5
	4 times a week	18	22.5	22.5	25.0
	3 times a week	29	36.2	36.2	61.2
	1- time a week	13	16.2	16.2	77.5
	Seldom	18	22.5	22.5	100.0
Total		80	100.0		

Table 1.1 summarizes the respondents' characteristics. There were 24 males (30%) and 56 females (70%), indigenous people from two separate indigenous communities. Table 1.2 shows the different categories of School Library (SL). There were 65 (80.2%) respondents from urban area, 13 (16%) respondents from rural area and only 2 (2.5%) respondents from special government school program or boarding school. Table 1.3 shows the frequency

respondents visited SL from more than 5 times to less than 1 time a week.

DISCUSSION

Table 2. The results of ANOVA

		Sum of Squares	df	Mean Square	F	Sig.
1.1 SL staff members are efficient in helping students to find resource materials	Between Groups	4.157	2	2.078	7.010	.002
	Within Groups	22.831	77	.297		
	Total	26.987	79			
1.2 SL provided fans/air-conditioners for comfortable studying atmosphere	Between Groups	2.357	2	1.178	5.391	.006
	Within Groups	16.831	77	.219		
	Total	19.188	79			
1.3 SL provided website for accessing reference materials	Between Groups	7.103	2	3.551	5.771	.005
	Within Groups	47.385	77	.615		
	Total	54.488	79			
1.4 SL environment are conducive for users	Between Groups	2.126	2	1.063	4.339	.016
	Within Groups	18.862	77	.245		
	Total	20.987	79			
1.5 Is it necessary to change the current SL Manager/Head with a full-time, professionally trained librarian?	Between Groups	2.934	2	1.467	4.070	.021
	Within Groups	27.754	77	.360		
	Total	30.687	79			

Note: α= 0.05; N= 80; SL (School Library)

Table 1.1 shows there is significant difference in SL staff members to helps users or students to find resource materials. Librarians or library teachers have limited time in the library while students felt apprehensive about communicating with them for help. This effect was more prevalent in rural students. Even though library teachers who came from urban areas or schools had training in library management, they could not give proper advice to students in finding the right resource materials due to limited knowledge in the field. On the other hand, full-time and professionally-trained librarians received specific training to effectively guide students in their research for appropriate materials. Table 1.2 shows that there is significant difference to provide the facilities for comfortable studying environment in SL. Also, there are differences between urban SL compared with rural SL. Conducive atmosphere contributes to maximum library usage. All SL systems in Malaysia received government support in terms of monetary and equipment funding. However, disparities exist in the type of supports received. For example, urban schools received more supports as in volumes of book collections, air-conditioners, fans, shelves, chairs, and tables compared to rural schools. All of these supports created an ergonomic environment for the library. Interestingly, boarding schools had advantages over rural and urban schools mainly due to donations from wealthy parents. Table 1.3 shows that there is significant difference in the location where online computer facilities available to search websites for accessing reference material. All government- supported schools have funding allocation to create their own SL. SL in rural areas provided less computer facilities compared to urban and boarding schools. Monetary funding from the government is limited and based on individual effort. Some schools may seek support from external agencies and parent-teachers associations to purchase computers for their SL. Internet access for the purpose of locating reference materials to some rural schools are very limited compared to more advanced schools in urban areas. Schools in Malaysia do not have paid- or password-access Internet services. However, users have access to free websites for searching reference materials. As a result, students need proper guide from professionally trained librarian; especially in guiding them to appropriate websites for reference and research purposes. Table 1.4 shows that there is significant difference in SL where environment are conducive for users. SL usage experiences are different in rural than urban and boarding schools. Qualities in services and facilities also played critical roles in creating pleasant environment for users and visitors. The volumes of books were more available in urban and boarding schools. Well managed and conducive SL requires adequate space for readings, discussion areas, comfortable chairs and tables, appropriately display shelves for manageable browsing, and ergonomic layout. All of these criteria are essential to get maximum use out of SL and to keep users and visitors interested and returning again. However, all of these are possible with adequate support and funding.

Table 1.5 shows that there is significant difference in leadership and managerial in SL compared to rural with urban and boarding school. Is it necessary to change the current SL Manager/Head with a full-time, professionally trained librarian?"--Education, formal training, and full-time commitment of professionally trained library in Information Management are critical factors in SL success. Teachers are focused more on classroom teaching and other academic responsibilities compared to librarians. Some teachers may receive proper training but have constraints to fully manage the libraries due to academic obligations in and out of the classrooms. Therefore, professionally-trained librarians are needed in Malaysia's SL, especially in secondary schools, to replace library teachers whose primary focus are and should be on teaching and learning

RECOMMENDATIONS AND CONCLUSIONS

Based on the findings, the researchers recommend the following that would be beneficial for the government, Ministry of Education, and administrators to ensure the successful implementation of meaningful library program and ergonomically designed school libraries:

- Malaysian government must formulate policies to ensure the person whom in charge in school libraries are from those who received a standard required qualification pertaining to school libraries of resource center especially graduate students from *Bachelor of Science in Information Studies (Hons), (Information Resource Centre Management), Universiti Teknologi Mara.* They also need to immediately update the education funding formula salary benchmarks, so that funding for school can be spent also on school libraries (The Ontario Library Association, 2006).

- Figure 2 shows school library and resource center with mobile book racks. The availability of mobile book racks and digital technologies mean for student and rural villagers is to encourage the selective and choice for various categories of student and rural villagers both male and female. The digital technologies that commission in the school libraries is another attraction not only for student during the school hours but also as a learning and communicative tool for rural villagers during the off hours and weekend, as not all students and rural villagers could afford to possess the state-of-the-art digital technology as learning tool. Even the e-book section could be fully utilized by the students and school teachers and staffs.

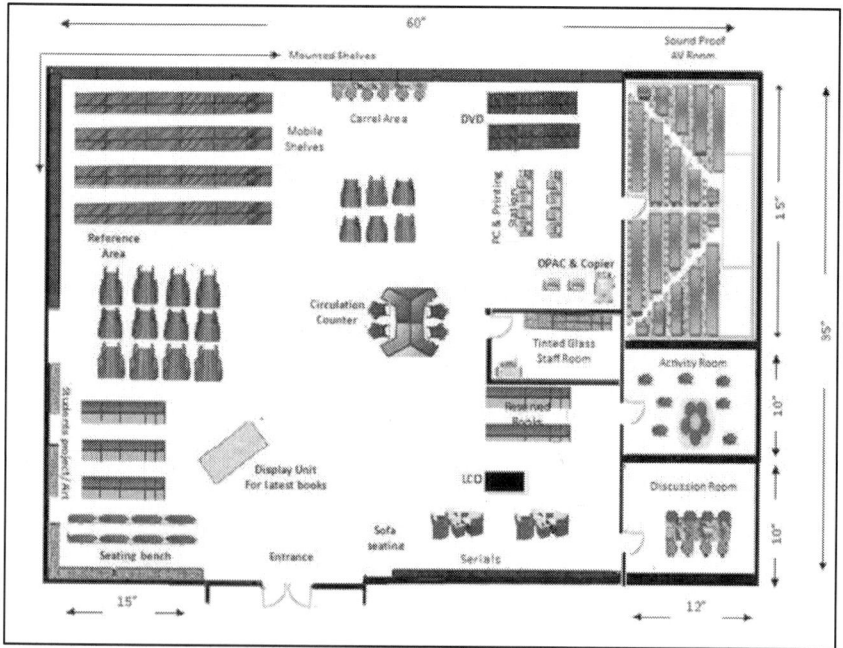

Figure 2. Proposed new floor layout for user-friendly mobile digitalized school library

- The retraining under in-service courses for all teacher librarians indeed will facilitate to fulfill the need of operating the new concept of mobile digitalize rural school libraries and resource centers. Teacher librarians should follow in-service training to gain latest state-of-the-art knowledge on the operating of digitalize mobile libraries in rural schools. According to Aminuddin (2011), organization particularly government institutions such as schools, did not see the need to train and develop their staff. However, in an era of economic transformation plan introduced by the current government, all public institutions have realized the importance of developing the capabilities and enhance the abilities of their employees in order to meet the increasing demands and expectations of the government and the public. Employee training and development is the organizational activity which aims to improve an employee's current performance, change attitude, or develop skills.

- Government through the Ministry of Education should draft out the ten years' modernization of rural school libraries plan and immediately implement it throughout the country because it is needed to create and build the country's human capital to further develop a knowledge and learned society.
- Rural school libraries should be upgraded and increased both hardcopies of books, digital technologies, and media. All libraries should be air-conditioned as the country is located in the tropical region and at times can be very warm and humid. This is to encourage students and villagers to frequently visit the rural school libraries to read, as reading habit is very important to every individual.
- Governmental task force should be formed to supervise and advise teacher librarians in term of fully utilize rural school libraries. Book volumes and collections in variety of topics must be supplied to the libraries.
- Based on the revolution that derived from the U.S National Commission on Libraries and Information Science in 2007, it has
- approved and affirmed a clear link between school library media programs and student achievement when those libraries are staffed by an experienced school library media specialist. The

physical location of the library and its infrastructure need to be redesigned. A library is not just a place with books, but as an integration center to ensure that all members of the school community will have fair access and learning opportunities to an open range of resources, either printed and non- printed, electronic media, and IT usage. As library can be remodeled and expanded over time, an appropriate physical location of a school library must be taken into account to determine suitable architecture planning and library design. Building the library near the high traffic area of the school will offer easy access for patrons and increase higher flows of visitors. Location of library at hidden area, higher floor or separated from the main school building are inconvenience for school community and public patrons especially for handicapped and disabled people. Therefore, the entrance should be designed according to patrons' requirements so as to certify maximum visibility of the school library. Further, systematic operating hours will allow school community and public patrons to use the library and its collections when they need to regardless of breaks, weekend or public holidays.

- An adequate space is crucial to provide ample facilities and great atmosphere for school library patrons. Small library space surroundings will make the library collection and equipment looks visually chaotic plus, limiting the number of patrons per visit. The basic furnishing equipment for a library is bookshelves. Books are easier to find if they are displayed on shelves. The height of bookcase at secondary school normally should not be higher than 180cm. In order to maximizing the space of a library, the use of mobile shelves is the best. This roller racking systems are designed to amplified the capacity of traditional storage shelves while eliminating uneven aisle between shelves. Thus, more floor space can be utilized for storage and it is easier to move. Great atmosphere and appealing interior design is one of the elements that affect directly to boost the number of occupancy and length of stay in a library. The design of the library should work with space, color furniture and functionality which will tie the relationship between patrons, library staff and its collections. For

seating arrangement, a variety and choice of desks and chairs should be offered. A lazy chair for instead, is not only a stylish and incredible comfortable armchair, but is an example of choices for teens in particular as an appreciation for them by providing a playful and fun seating area. In addition, modern school libraries were suggested to have an open and close reading or discussion area. This is provided within and outside the library. The open reading area outside the library must be equipped with automatic or retractable awnings. It will create the perfect complement to the outside library decor and providing protection from the rain and sun. Retractable awnings tend to have a longer life because they can be retracted during severe weather.

- Establishment of the school library requires an understanding about its importance by appointing well qualified personnel. A broad librarianship skill is the main requirement for the person in charge carries out through professional training. As most of the libraries are headed by teachers, to effectively coordinate the role of the library in the school, this staffing should come from a professional librarian or a dually qualified teacher-librarian. This will raise the standard of facilities and efficiently serve the patrons' needs. Professionals have depth knowledge in preparing the library room and organizing its collection and equipment. They also know on how to operate and maintain the arrangement of information by subjects for example or standard classifications like DDC or LCC. Including library catalogues, shelf list and title catalogue which are not easy to do by non-professionals. Moreover, professionals are able to assist patrons using exact information retrieval skills and guide them to find books easily. Rare collections, old manuscripts, collections on public figures, nation leaders, teaching icon and etc. are proficiently indexed and archived by them.

- Bank of computer terminals, servers, digital learning resources range from lesson plans, images, audio, video, to interactive e-learning materials should be provided in every school library. The networked computers linked the library from classrooms and labs

enabling access to library resources, licensed databases and the Internet also raised by respondents as definite tools and mediums to share information and knowledge between students, teachers, and local community or public patrons.

Digitized materials in the form of CD and through the World Wide Web will encourage patrons to seek for useful information like past years' exam papers, online learning, parenting and other useful knowledge to be used in their daily routine.

Library and resource center in rural schools should provide benefits to all, not only to cater for the needs of students, but also to contribute to the good cause to instill rural villagers' reading habit during the off hours of school session and weekend. In order to attract not only the students and villagers to utilize the school library, both during school hours and off school hours respectively, a new featured and model of state-of-art mobile digitalized school library need to be drafted and proposed to the government. The two prong usage of rural school user-friendly library to cater the need of students and villagers. Consequently, the main aim is to build human capital of learned and knowledged Malaysian society, as well as to realize 2020's Vision of Malaysia to become a developed country where Malaysians from all walks of life, both young and old, can be called Learned Society.

REFERENCES

1. Aminuddin, M. (2011). Human resource management principles and practices (2nd ed.). Kuala Lumpur, Malaysia: Oxford University Press.

2. Facility. (2011). Retrieved September, 2011 from <http://dictionary.reference.com/browse/facility>

3. Hair, J. F., Black, B., Babin, B., Anderson, R. E., & Tatham, R. L. (2006). Multivariate data analysis (6th ed.). Upper Saddle River, N.J. Pearson Prentice Hall.

4. Jusoh, F. (2002). School libraries in Malaysia. Paper presented at the 2002 IASL Conference.

5. Petaling Jaya, Malaysia. Retrieved from http://www.iasl-online.org/events/conf/conference2002-fatimah.html

6. Kamal, M. A., & Othman, N. (2012). Training and development for library and media teachers in selected Malaysian school resource centres. *Journal of Education and Practice*,3(6).

7. Lance, K. C. (Winter, 2004). Libraries and student achievement: The importance of school libraries for improving student test scores, Threshold, Retrieved from in http://www.ciconline.com.

8. Lance, K. C., Rodney, M. J., & Hamilton-Pennell, C. (2005). Powerful libraries make powerful learners: The Illinois study. Canton, IL: Illinois School Library Media Association.

9. Laporan pelaksanaan Pusat Sumber Sekolah (PSS) Tahun 2001 (2001). (Unpublished), Kuala Lumpur: Bahagian Teknologi Pendidikan, Kementerian Pendidikan Malaysia.

10. Raja Yaacob, R. A, & Omar, S. (2003). the training of teacher librarians in comparison with professional librarians in Malaysia. *Malaysian Journal of Library & Information Science*, 2, 27-41.

11. Scholastic Library Publishing. (2008). Research foundation paper: School libraries work (3rd ed.). Retrieved from http://www.scholastic.com/content/collateral_resources/pdf/s/slw3_2008.pdf

12. School Library. (2011). Retrieved from http://www.wbdg.org/design/school_library.php

13. Singh, J. (2005). Challenges faced by Bank Negara Malaysia as a knowledge based organisation to move knowledge management. (Unpublished doctoral dissertation). University of Melbourne, Australia.

14. The Ontario Library Association. (2006). School libraries and student achievement in Ontario. http://www.accessola.com/data/6/rec_docs/137_eqao_pfe_study_2006.pdf.

Chapter 1

Towards a better Design: Physical Interior Environments of Public Libraries in Peninsular Malaysia

Suhaila Sufar[a*]**, Anuar Talib**[b] **& Haris Hambali**[c]

[a]*Dept. of Built Environment (AP780) Faculty of Architecture, Planning & Surveying, UiTM Shah Alam* [b]*Dept. of Interior Architecture, Faculty of Architecture, Planning & Surveying, UiTM Shah Alam.* [c]*Dept. of Interior Architecture, Faculty of Architecture, Planning & Surveying, UiTM Perak*

ABSTRACT

The paper will focus on how to create a better design of physical interior environments in public libraries in Malaysia. The research will consolidate on the interior ambiences caused by selection of lightings, furniture, materials and finishes of public libraries. It will discuss on the previous studies and literature reviews which aimed to evaluate and review their opinions on the influence of physical interior environments on users' expectations, needs and behaviors on the interior settings. This overview will help the researcher in future study to investigate on the interplay between library users and physical environment, particularly on spatial ambience and the users' behaviors.

INTRODUCTION

The public library is an importance as community assets. The subject that is being endlessly debated is about the future of libraries. Nowadays it is a place of knowledge acquisition, communication and socializing with other people. Public libraries and university libraries start to change the "rusty" approach of a quiet place to a new modern place of meeting other people and using information in groups. We can also see how the architecture and interior in libraries has changed during the time. New shapes, colors and concepts are employed in newest projects and interior solutions in different libraries all around the world. According to Juhnevica & Udre (2010), the needs of users, interior space planning and physical interior environment design should be at the centre of the planning process. Future of libraries should serve multiple roles communities and society. However, the success of the design will be achieved by the user's satisfactions on how the physical interior environments influence them. Latimer, K. (2007), points out that, user space needs to be well planned, welcoming and attractive as the role of the physical library and the needs of users change in the 21st century. The importance and interest of the study is to show that such issues as interior design and architectural solutions can be important in creating appropriate and friendly place for library users.

BACKGROUND

According to UNESCO (1999), shows that the report of reading habit among Malaysians were only read a couple of books a year. This report shows that, we are far left behind contrast with other developing countries. This obviously shows that the reading habits among Malaysians was still low and need to be generated with more consolidation for improving and in producing knowledgeable society and community. When it comes to the public library scenario in Malaysia, it is very different and far out to libraries in developed countries. This includes the interior design quality, comfort level, satisfaction, customer experience and the characteristics of the physical design of the interior environment. In researcher opinions, now is the time to address the issues in lack of reading habit among Malaysian and to recognize the reason why Malaysian people not interested to visit libraries. It may be related to less attraction to visit, lack of facilities infrastructure and less comfortable in interior space.

UNESCO's statistics (2009) also placed Malaysia at 11th (of 137 nations) that allocates a huge budget for national education. In the effort to create a society with knowledge, not only formal education comes into focus, instead inculcating reading habits is among the core values of a society with knowledge. Here, the role of the traditional libraries as bookstores of information, a quiet and bored place to visit, now becomes a place of acquiring knowledge: through socialization, communication, in a relaxed interior space. Despite the developments of accessible information and the "online" technology today, the physical library is still important as a public building that facilitates the acquisition of knowledge.

In some countries the library may be the only physical space for learning and sharing information in a non-institutional context. Other societies offer a wide range of possibilities and it becomes crucial for libraries to deal with different partners in lifelong learning and education. High-tech learning centres have the potential to become attractive working environments for the internet generation when they integrate technology in an appealing overall picture with communal and social spaces. The open learning space at last can serve the knowledge societies by offering low intensive and collaborative meeting places for their communities. The ideal way would be a comprehensive, multifunctional space combining all three approaches beneath the same roof. A library built and organized with this in mind could become an attractive place for informational participation and lifelong self-paced learning. (Eigenbrodt,2009)

Furthermore, the development of library building concepts should be evolutionary with new inspiring design and features appearing as the changing needs of the people. Library design is not only about the exterior envelope but is also about practical and exciting physical interior spaces and environments. One of the most important aspects for designing a library is to consider the physical interior environments which comprises of interior space planning, and interior ambience such as selection of lightings, furniture, materials and finishes.

Suhaila Sufar et al. / Procedia - Social and Behavioral Sciences 42 (2012) 131 – 143

The research questions of this research are:

- What is the physical interior environment should be considered in designing public libraries?
- What are user's expectations on physical interior environments in public libraries?
- How well the physical aspects and design of the public library had catered to the needs and expectation of users?
- What are the key criteria to achieve a better design in physical interior environments in public libraries of today and future?

The objectives of the research are:

- To study and explore on the various functions and related of physical interior environments in public libraries.
- To study the factors that related to the user's expectation on physical interior environments in public libraries.
- To study on user's perception and behaviors towards the physical settings of interior environments in libraries

Significance of the study is:

Firstly this study evaluates the physical interior environments through expert views, user's perception and their needs. Secondly, the study will improving the existing guideline and information which are produced before by the National Public Library, Malaysia (NPN) and Malaysia Public Works Department (JKR) in order to establish a key criteria of physical interior environment to achieve a better design of public libraries in the future. The study also will be based on evidence base design; which study on user's behaviors towards the ambience of interior environments.

LITERATURE REVIEW

This paper will begin with a review of the recent literatures in the field of library building design. The discussions will be consolidate on the definitions of public library, what are the criteria of physical environments which regards to the interior ambience such as various functions and importance of physical interior environments in public library, and the impact of the interior ambience and physical interior environments which influence human behaviors and user's experience.

Definition of Public Library

Generally, library is the central sources of knowledge and information have the same basic objective of providing services to disseminate knowledge and provide reference materials to customers who need the service. Libraries in the digital age of today, not only physically as a space for knowledge, but also a place for individuals or groups come together to explore, learn, meet, interact, socialize as well as enjoy democratic access to resources and information. Therefore, providing a comfortable space, meet the taste and satisfaction of interior space and physical environment in general is very important to attract users to come and use the library services.

According to UNESCO (United Nations Educational, Scientific and Cultural Organization), a public library is defined as the local centre of information, making all kinds of knowledge and information readily available to its users. The services of the public library are provided on the basis of equality of access for all, regardless of age, race, sex, religion, nationality, language or social status. Specific services and materials must be provided for those who cannot, for whatever reason, use the regular services and materials, for example linguistic minorities, people with disabilities or people in hospital or prison. All age groups must find material relevant to their needs. Collections and services have to include all types of appropriate media and modern technologies as well as traditional materials. High quality and relevance to local needs and conditions are fundamental. Material must reflect current trends and the evolution of society, as well as the memory of human endeavor and imagination. Collections and services should not be subject to any form of ideological, political or religious censorship, nor commercial pressure.

According to IFLA (Institute Federal Library Association, (2007), a public library is an organization established, supported and funded by the community, either through local, regional or national government or through some other form of community organization. It provides access to knowledge, information and works of the imagination through a range of resources and services and is equally available to all members of the community regardless of race, nationality, age, gender, religion, language, disability, economic and employment status and educational attainment.

McCabe & Kennedy (2003), add that the public library is a place that belongs to everyone. New libraries often are arranged like bookstores, and many offer light refreshments that encourage customers to linger. Libraries are increasingly becoming places for interaction, and most new ones have meeting spaces, seminar rooms, and sometimes performing- arts spaces and galleries. All of these activities and attractions make the library a desired destination and increase the square footage needs for new libraries.

According to Rooney-Browne (2009), she points out that the public library spaces should transform into vibrant, welcoming destinations; exploiting new revenue streams;attempting to establish spaces in virtual communities; and with the help of a SWOT analysis. SWOT analysis is refer to S-strenght, W- weakness, O-opportunity, T-threat.

Physical interior environment

Physical interior environment are one of the major issue that have been continuously debated and discussed in order to create a better design and attraction to the users. McCabe & Kennedy (2003), point out that, a library building must be attractive and aesthetically pleasing to the eye. Internally, it must be functional and current in the use of technology supporting services to its community. Architectural design features provide for the physical appearance. The descriptions that follow will appear to meld issues of attractiveness and effectiveness, because sometimes they are inseparable. The descriptions are applicable both to new buildings and to buildings requiring renovation.

The physical environment encompasses both architectural elements such as physical layout, furniture, and equipment and visual sensory elements such as color, texture, and lighting. These two aspects, in conjunction with ambient factors, create the interior environment. (Bitner M. J., 1990). In an upscale restaurant, for example, patrons expect well-prepared food and attentive service, but they also expect comfortable seating, mood-setting lighting, pleasant or luxurious décor, soothing music, and an opportunity for social interaction. In these situations, the emotional components of the service setting become more important and can strongly influence the consumer's ultimate assessment of the quality of the service as a whole. (Jiang, 2006).

According to (Anandasivam & Cheong, 2008), the biggest challenge is to gain more teenagers and young adults to the physical library. The library needs to become "cool" and comfortable so that teenagers could consider it as a good place for meetings and hanging out together. Young people prefer comfortable workplaces and more freedom to move around and explore the space; they need a place to use their laptops and different zones to work. Many young people prefer to work in open space workplace together with others but still some of them want to work in silent rooms. There is also a big need of silent rooms in the library, cited in (Juhnevica & Udre, 2010).

Furthermore, the libraries need to have all technological equipment to make working in the library as easy as possible (Childs, 2006).Users need to have approach to printers, scanners and electronic catalogues. Also the planning of the space in the library is very important so that users would not feel lost. The best approach is to plan all storeys in the same way. There should be different zones for different purposes – reading, web browsing, group works, references and recreation. There are many interesting ways how to make library building more comfortable and cozy, (Juhnevica & Udre, 2010).

Physical interior environment criteria's and parameters

In designing a better way of physical interior environments in public libraries, there are several criteria and parameters that need to be taken into account. The physical environment comprises of elements of interior such as physical layout, selection of furniture, and equipment and visual sensory elements such as color, texture, materials, finishes and lighting. These two aspects, in concurrence with ambient factors and creating the interior environment. On the other hand there are modern libraries which have a lot of open space, nice and comfortable furniture like sofas and armchairs, coffee-tables and many computerized places or workspaces for users to work with their laptops. Most of the libraries have free Wi-Fi Internet access, electronic catalogues and other technical equipment to make library functions more available for a greater range of users. It all matters when we look at latest trends in library interior and design. (Juhnevica & Udre, 2010). All senses can be involved – music, nice but not annoying smell, good lightening, comfortable furniture, pictures on the walls and many other things can make users feel welcomed. Variety of different furniture, different zones and places to gather or be alone can provide different types of users with just what they need. The library might become a very good place for exhibiting art and at the same time it could create a special atmosphere and make the library more dynamic (Anandasivam & Cheong, 2008).

Color selection

To begin, as McCabe, G. B. (2003) puts it, should study on the psychological effects of color. Dark colors may subdue excitable behavior, bright color will stimulate. According to Gold Coast City Council Branch Libraries (GCCC), 2007, a contemporary approach to color and materials selection should be adopted so as to evoke an inviting and friendly image with a sense of efficiency, coupled with a stimulating memorable building experience and civic presence. The new library building should demonstrate a clean line aesthetic, backed up by a layered, textural and timeless approach to the colour and materials palate. Colour should be incorporated selectively and based on a neutral 'background'. Colour may feature on elements such as selected walls and partitions, the fabric selection for furniture and the use of graphic elements to clearly define distinct areas of the library. Colours and finishes should be appropriately selected to reinforce the appeal of the spaces to the designated user age group, while seamlessly meshing with the total concept of the space as an inviting, stimulating place to visit. Nowadays there are more bright colours in the libraries.

Furniture and equipment

Variety of comfortable furniture and different furniture, different zones and places to gather or be alone can provide different types of users with just what they need. (Anandasivam & Cheong, 2008).

Standard Shelving -The consultant and the librarian will calculate the need in linear feet for bound volumes with an allowance for growth. This figure will be converted into shelving units, single or double faced, and in heights ranging from 42", or about 60" to 66", to full 84" and 90" heights. The shelving preferably should be steel. Some librarians may prefer the lower height shelving choices with very few units, if any, above the 66" height. The preferred unit has the base on the floor with shelves above. T- based shelving isn't recommended for two reasons: losing the lower shelf and the difficulty in cleaning under them. If the ethnic make-upof the area population tends to be people of shorter stature, then the lower Furnishing and Equipping the Library and Its Environs 45 shelving may prove more satisfactory. Capacity in volumes will be lower because of fewer shelves. (McCabe G. B., 2003). According to (Gold Coast City Council Branch Libraries (GCCC), 2007), The maximum height of book-stacks is limited to 1500mm above the finished floor to reinforce the fit-out transparency and to be in line with OH&S and Equitable Access guidelines for staff and customer use.

Seating- Seating space is a crucial element in library design. Most public and academic libraries find that space to seat the growing user population is increasingly encroached upon by shelving to house growing collection. Library users should have a variety of seating to choose from to match their learning approach and mood at a particular given time there are essentially three types of seating: 1) Reading benches,2) Reading tables,

3) Flexible seating. Based on the user feedback on seating, they find it convenient to form informal discussion spaces with the movable seats. Students are attracted to this area by the striking color and "cool" atmosphere. (Anandasivam & Cheong, 2008)

Lighting

Lighting can control how the library looks and feels. Good lighting is needed for study, but intimate lighting may be excellent for reflective areas. Use spot lighting for special effects. Planners of both new academic and public library buildings are also concerned with the external appearance of their facilities. In his paper on Flushing Library in New York, Gary Strong praises the curtain wall façade of the building which "allows the public to see library activity within and invites them into the building" (Strong,2001:127). This concept of transparency remains a popular feature in libraries today, such as the Lanchester Library (Noon, 2004) and the Idea Store in Tower Hamlets, where "large amounts of glazing to enable local people to see what was going on inside were essential" (Wills, 2004:109). At the Hatfield Campus Learning Resource Centre the designers took care to build the glass façade along the northern side so as to reduce solar glare problems for the readers inside (Martin, 2004). However, McCabe (2000) outlines the various issues that arise from using a large amount of exterior windows in a library, including security and safety as well as avoiding glare and overheating."The earth has its limits, and buildings need to be environmentally friendly as well as handsome and cost-efficient." (Sannwald, 2007:135) The IC (University of Sheffield, 2007a) is of course one example of this – another is the Lanchester Library (Noon, 2004) which uses lightwells and natural ventilation to reduce the need for artificial light and air conditioning, and has windows carefully aligned to minimize solar gain and glare.

(Pierce 1980, 30) support that, for many years the common understanding when selecting reading surfaces has been to avoid colors or backgrounds that are either too light or too dark. In recent years black and other very dark reading surfaces on tables and carrels have appeared. Why this change? This question is worth investigating, because some serious issues are involved. William Pierce, a well-known consultant, cautions against the extremes of light and dark surfaces as well as glare: "The specifies should be cautioned about using top finishes that are too light or too dark. The too-light top is uncomfortable to the eye of the user and difficult to keep clean; the very dark top provides too great a contrast to paper; thus is uncomfortable to the eyes".

According to GCCC (2007), lighting design in library should be considere on 1) Make lighting glare- free with a minimum level of 50 lux at ground level. 2) Eliminate glare, illuminate signage and highlight level changes. 3) Provide uniform illuminance levels internally and comply with the requirements for maintenance illumination in all circulation spaces, including publicly accessible areas. 4) Provide a minimum illumination of 40 lux, uniformity of 0.3 and an average maintained value of 120 lux. Provide a graduated level of illumination at building entries and exits to assist people with vision impairment. Provide a minimum of 50 lux outside the entry or exit.(5) Provide adequate focused lighting for sign language interpretation for people who are deaf/hearing impaired in conference rooms, meeting rooms, auditoria and the like.

Ambient and psychological aspects of the physical interior environment – affect the human behaviours and perceptions.

It also should also consider on how elements of the physical environment influence human attitudes and user's behaviors. According to Shill & Tonner (2004), points out that some previous studies and surveys on usage and library design have indicated that the design of the physical environment can have significant impact on library usage. Research suggests that the physical setting may also influence the customer's ultimate satisfaction with the service, productivity, and motivation (Bitner 1990;Harrell,Hut, and Anderson 1980). According to Darly and Gilbert 1985;Holahan 1986; Russell and Ward 1982; Stokols and Altman 1987) in Bitner, M. J. (1992), add that human behaviour is influenced by the physical setting and interior environment in which occurs is essentially a truism. Interestingly, the field of environmental psychology has addressed the relationship between human beings and their built environment. Here, it is assumed that dimensions of the organization's physical surroundings influence important customer and employee behaviors. According to (Mehrabian and Russell 1974) cited in Bitner,

M. J. (1992), that environmental psychologists suggest that individuals react to places with two general, and opposite, forms of behavior;approach and avoidance. Approach behaviors include all positive behaviors that might be directed at a particular place, such as desire to stay, explore, work and affiliate. Avoidance behaviors reflects the opposite, in other words, a desire not to stay, explore, work, and affiliate. (Mehrabian and Russell 1974). In other words, the approach of user's behaviors in interior space (including enjoyment, exciting, exploration, attractions, stay longer and returning) are depends on the physical setting in interior environments of libraries.

According to (Bell, 2008) the literature shows that the physical environment influences customer behaviour by:

- Creating strong emotions and mood by means of the sensory qualities of the environment (Grossbart et

- al. 1990; Baker and Cameron 1996; Sherman, Mather, and Smith 1997; Richins 1997; Le Bel 2005; Berry, Wall, and Carbone 2006);

- Influencing personal interactions between patrons and staff and among patrons (Le Bel 2005);

- Creating a pleasurable total customer experience (Berry, Carbone, and Haeckel 2002); and

- Influencing future patronage intentions (Carbone and Haeckel 1994; Ward and Bitner 1992; Berry, Carbone, and Haeckel 2002; Berry, Wall, and Carbone 2006).

Physical environment includes sensory aspects and has the power to affect mood and emotion. Perception of the environment, in its most strict sense, refers to the process of becoming aware of a space by the acquisition of information through the sensations of sight, hearing, smell, touch, and taste. Cognition is the mental processing of this sensory information. This may involve the activities of thinking about, remembering, or evaluating the information. Spatial behaviour refers to responses and reactions to the environmental information acquired through perception and cognition. The designer creates environmental stimuli to direct these psychological stages as well as the secondary processes of motivation, effect and development. Environmental expectations, another determining element to be considered by the interior designer, are developed over time through experience and interaction with the environment. Sensations, in combination with expectations of the environment, define one's perception of a space. The influence of atmospheric attributes in marketing contexts is based on the premise that the design of an environment through a variety of means such as temperature, sounds, layout, lighting, and colors can stimulate perceptual and emotional responses in consumers and affect their behavior. (Yildirim, Akalin-Baskaya, & Hidayetoglu, 2007).

Ambient influences

According to (Bell, 2008) cited that ambient influences include the non-visual sensory or atmospheric aspects of a service setting, including sounds, smells, and temperature. The literature clearly reveals that these elements have a strong influence in the perception of service quality in hedonic consumption settings (Hirschman and Holbrook 1982; Reimer and Kuehn 2005; Jiang and Wang 2006). Hedonic consumption is based on appeal to the senses and the quality of the sensory input is directly correlated with the perception of service quality. Because these atmospheric influences can arouse intense emotions and create mood, they strongly contribute to the total customer experience (Carbone and Haeckel 1994; Le Bel 2005).

According to Bitner (1990), she points out that, although ambient influences are important in hedonic situations, they are only weakly influential in utilitarian settings. Because utilitarian consumption experiences are based on functionality and sensory influences do not typically affect functionality, the use of sensory stimuli in a utilitarian setting is at best peripheral to an assessment of service quality. Ambient music, for example, can be used to create a pleasant background in a utilitarian service setting, but this element would not be likely to strongly factor into a customer's patronage decision. Ambient influences work in tandem with the visual aspects of the physical setting to create a complete interior environment Ambient stimuli, by their ability to arouse emotions, are a major influence on the psychological aspects of the service setting and can provide strong clues to service quality (Bebko, Scuilli, and Garg 2006; Berry, Wall, and Carbone 2006).

Psychological influences

Psychological influences include emotion, mood, and attitude. Reference to emotions and mood permeate the service quality literature and customer satisfaction and perception of service quality are strongly affected by consumers' psychological reactions (Gardner 1985; Hirschman and Holbrook 1982; Reimer and Kuehn 2005; Jiang and Wang 2006). The interior environment has a powerful ability to generate emotions and those emotions influence customers' perceptions of service quality (Baker and Cameron 1996; Sherman, Mather, and Smith 1997; Richins 1997; Le Bel 2005; Berry, Wall, and Carbone 2006). This is particularly true in hedonic service settings. Although emotions can have an effect on utilitarian encounters, this influence is weak. Consumers' psychological reactions are closely related to the ambient input available at the service site and the physical qualities of the service environment. Psychological influences are also related to the sense of time and to social interactions.

Important role of public library as a social space

According to (Bell, 2008) he cited that likewise, today's library encourages more social interactions within the library and offers a range of group and children space. These changes in library use have forced librarians to rethink their approach to the planning and design of the library building. Many writers emphasize the important role that the library (particularly the public library) plays as a social space, and how space planning must reflect this. Lang (1999) argues that "libraries throughout history have been extraordinarily successful socialspaces" (11), and others agree that the library can help to satisfy a "yearning for informal socialization" (Lushington, 2002:91) and is important as "a public space and meeting place" (Dewe, 2006:21).

The increasing importance of this role, particularly for public libraries, usually results in both open- plan layouts in the main part of the library that maximise contact between different elements of the community and also separate facilities like cafes and meeting rooms for community groups which can help preserve a quiet atmosphere in other parts of the building (Dewe, 2006; Greenhalgh et al., 1995; Lushington, 2002; Sannwald, 2007). In his account of the development of the Lanchester Library Noon stresses that they were creating "a kind of shopping mall where the main attraction was... the Library!" (Noon, 2004:94).

Furthermore, nowadays public libraries becoming learning space and at the same time becoming a place where users come to acquire knowledge, communicate, socialize and take a rest. The new ideas of adding a place for refreshment such as a café or small coffee bar in the public libraries are becoming an issue. Some young people would like to visit the library as they are visiting coffee-shops – have relaxed atmosphere where they can chat, study or have a rest (Pomerantz & Marchionini, 2007).However, as Hurt (2000) admits, this means going against beliefs that are deeply held by many librarians, though she was glad that at George Mason University they "broke some rules that should no longer be automatically accepted" (103), including allowing food into the library. According to Dewe (2006), these and other taboos are being increasingly challenged in 11 public libraries too as "rules about eating and drinking have been relaxed and noise disregarded" (23).

Understanding on the effects of the physical environment and its relation in producing a better interior ambiances design

Understanding on the effects of the physical environment and its relation in producing a better interior ambiances design can be found in *Figure:1*. Physical interior environment is one of the most important aspects in enhancing better design in libraries. The researcher found that there are four physical interior environments should be emphasized in providing a better design in libraries such as that the selection of

- Lighting - day lighting (natural lighting) or artificial lighting. The selection of lighting is based on a

- standard comfortable lux and luminance.

- Furniture - selection of furniture that is appropriate according to the ergonomics of children, adults or a senior citizen. It also should be according to the needs of the group or individual user, a selection of furniture such as types of seating, reading tables, bookshelves and it also must be ergonomic and appropriate to each user.

- Space planning – the interior layout and space requirements,

- Materials and finishes - choice of colours and finishes are appropriate to the furniture, wall, floor and ceiling.

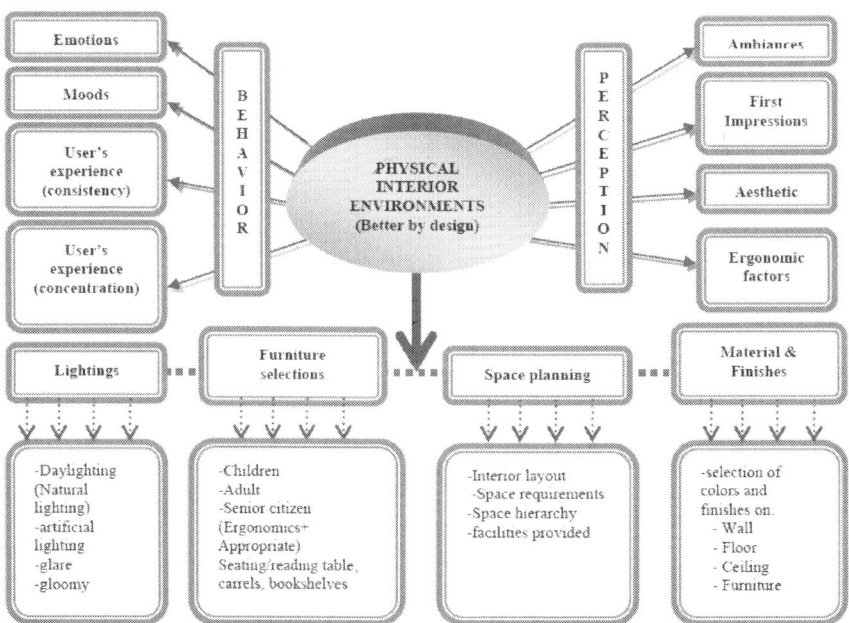

Figure 1. Understanding on the effects of the physical environment and its relation in producing a better interior ambiances design.

Table 1. Comparison of old and new approaches of the physical library interior aspects. Source: (Juhnevica & Udre, 2010)

Criteria	Old libraries	Current/digital age libraries
Concept of the library	Silent place for studying and working	Active place for retrieving information and socializing with other users
Space in the library	Much closed space, furniture close to each other	A lot of open space, space to move around between furniture
Rooms in the library	Big silent reading rooms, separated computer rooms	Big active reading rooms to use personal laptops with free Wi-Fi Internet access, and socialize with other users; separated silent reading rooms; additional computers in the library; special rooms for group-works
Lighting	Dark reading rooms with little lamps and extra lamps on reading desks	Bright reading rooms with a lot of light and extra lamps on reading desks. Design features where appropriate are the use of atria and skylights and using glass windows for maximizing the use of natural lightings.
Colours	Dark colours (brown, gray, beige, gold, deep red, deep green)	Bright colours (orange, blue, yellow, green, white, silver)
Furniture material	Heavy wood, chipboard	Plastic, textiles
Desks	Heavy, immovable, made of wood	Light, easy to move, made from different materials
Chairs	Heavy and hard wooden chairs, may be with leather seats	Easy portable and comfortable chairs with soft padding
Shelving	Big and heavy shelves, very tall ones with additional ladders	Easy-to-reach shelves, movable shelves, automatic shelves

The researcher also found that there are some aspects that affect emotions, mood and user's experiences. It also influenced human behaviors and perceptions through aspects of ambient, aesthetics and ergonomic factors. Those aspects affect users to come back to the library, to stay longer and find that the library is a fun and exciting place to explore and visits.

There are also some comparisons of old and new approaches of the physical library interior aspects of past and present, or the digital age. Summary of the characteristics of the old and new of the library interior are included in *Table 1*. The comparison is carried out by using observation of the libraries and doing source research (Anandasivam, K. & Cheong, C. F. 2008; Childs, P. 2006; Hulse, P. 2007;

Pomerantz, J. & Marchionini, G. 2007; Tseng, S. 2008) cited by Juhnevica & Udre, (2010). It can be described with the research finding studied by several criteria to be considered, include the concept of the library, the interior space, lighting, colors, finishes and furniture selection.

CONCLUSION

It is important to understanding of the role of the physical interior environment on users' perceptions of creating a better design. In researcher opinions and understanding, to achieve a better design of public libraries, aspects of physical and environmental criteria should be emphasized from the beginning of the planning space.

By understanding all of the influences on the perception and expectation of users on creating a better design, decision makers can ensure that their approaches to those ambient and physical elements within their expertise appropriately interact with and enhance the other influences on the total user's experience and behaviors. It is more than just the functionality and aesthetics of the ambient and physical design, but also temporal affects, social interaction and psychological consequences. Therefore, to create and to achieve a better design for future public libraries, the various aspects of physical interior environments and interior ambience should be consider such space planning, selection of furniture, lightings, material and finishes. According to (Bell, 2008), the interior environment comprises two of these influences – physical and ambient – and strongly affects the remaining functional, temporal, emotional, indicative, and social influences. Each of these influences affects customer response both directly and indirectly through its affects on the other influences. The physical setting is one of the most prominent and pervasive of these influences. In many ways, this statement sums up the service quality literature as it has evolved to its present state. Although physical environments comprises many factors, including the ambient, interior environment, that interact and influence each other in numerous ways, ultimately, the focus must be on the users – his or her expectations, desires, needs, and experience.

REFERENCES

1. (PNM), P. N. (2001). *Garis Panduan Perpustakaan Khusus*. Kuala Lumpur, Malaysia: Ampang Press Sdn. Bhd.

2. Anandasivam, K., & Cheong, F. C. (2008). Designing a Creative Learning Environment: NTU's New Art, Design & Media Library. *The Electronic Library Oxford*, vol. 26, page 650.

3. Bell, C. R. (2008). *The Role of the Interior Environment in the Perception of Service Quality:A Business Perspective*. Marymount University, Faculty of the School of Arts and Sciences, Arlington, VA.

4. Bitner, M. J. (1990). Evaluating service encounters: the effects of physical surroundings. *and employee responses. Journal of Marketing*, 54 (2): 69.

5. Bitner, M. J. (1992). Servicescapes: the impact of physical surroundings on customers andemployees. . *Journal of Marketing*, 56 (2): 57.

6. Clemons, S. A. (2009). University Libraries as Third Place. *IFLA World Library & Information Congress:75th IFLA General Conference & Assembly*. Torino, Milan, Italy.

7. Dewe, M. (2006). *Planning public library buildings : concepts and issues for the librarian*. Hants, England: Ashgate Publishing Limited.

8. Economic Planning Unit & Costs, E. P. (2008 Edition). Guidelines and Rules for the Planning and Building Standards Committee . In *SECTION G-Library*. Putrajaya, Malaysia.

9. Eigenbrodt, O. (2009). Libraries and Learning Centres; Current Approach. *IFLA: Section on Library Buildings and Equipment* .

10.   Eigenbrodt, O. (August 2009). Physical Space in Theory and Beyond: Building Libraries for the Knowledge Socities. *World Library and Information Congress:75th IFLA General Conference and Assembly.* Milan, Italy.

11.   Environment), C. (. (2003, August). *Better Public Libraries* . Retrieved from www.cabe.org.uk/files/better-public-libraries.pdf . Francine, M. (2009). Theme: Libraries as Place and Space ,In the Words of the Users: The role of the urban public library as place.

12.   *World Library and Information Congress: 75th IFLA General Conference Assembly.* Torino,Milan, Italy.

13.   Gold Coast City Council Branch Libraries (GCCC). (2007). *Generic Design Brief for Gold Coast City Council Branch Libraries (GCCC).* Gold Coast, Australia.

14.   IFLA. (2009). Libraries and Learning Centre; Current Approaches. *News 2009 No. 1 Section on Library Buildings & Equipment* . IFLA. (10-14 August 2008). Library as space and place: theoritical approach. *World Library and Information Congress:74th IFLA General Conference and Council.* Quebec, Canada.

15.   Jiang, Y. a. (2006). The impact of affect on service quality and satisfaction: the moderation of service contexts. *Journal of Services Marketing* , 20 (4): 211.

16.   Juhnevica, E., & Udre, D. (2010). "Libraries Designed for Users" Nowadays Concept of Library Architecture and Interior. *BOBCATSSS 2010 @ Parma, Italy* .

17.   Khan, A. (2009). *Better by Design: An Introduction to Planning and Designing a New Library Building.* London: Facet Publishing. Kingsbury, A. (2007). *To What Extent Do New Library Buildings Meet The Needs of Users and Staff?* THE UNIVERSITY OF SHEFFIELD, Master of Arts in Librarianship.

18.   Lamar, V. (1987). Toward the Environment Design of Library Buildings. *Library Trends* , 361-376.

19. Latimer, K. (2007). *USERS AND PUBLIC SPACE: WHAT TO CONSIDER WHEN PLANNING LIBRARY SPACE*. Berlin, New

20. York (Walter de Gruyter – K. G. Saur): IFLA Library Building Guidelines: Developments & Reflections.

21. Matthews, G. W. (19-21 August 2009). Libraries as space & place:The effective use of space in university and public libraries; current issue. *IFLA World Library & Information Congress:75th IFLA General Conference & Assembly*. Torino, Milan, Italy.

22. McCabe, G. B. (2003). *Planning the Modern Public Library Building*. London: The Libarries Unlimited Library Management Collection, Greenwood Publishing Group.

23. McCabe, G. B., & Kennedy, J. R. (2003). *Planning the Modern Public Library Building*. London: The Libraries Unlimited Library Management Collection.

24. McCabe, G. (2000). *Planning for A New Generation of Public Library Buildings*. London: Greenwood Press.

25. Olaf, E. (2009). Physical Space in Theory and Beyond: Building Libraries for the Knowledge Societies. *IFLA World Library & Information Congress:75th IFLA General Conference & Assembly*. Torino, Milan, Italy

26. Olaf, E. (2009). Physical Space in Theory and Beyond: Building Libraries in the Knowledge Societies. *IFLA World Library & Information Congress:75th IFLA General Conference & Assembly*. Torino, Milan, Italy.

27. Rooney-Browne, C. (2009). Libraries as space and place (Public Libraries as impartial spaces in the 21th Century). *IFLA World Library & Information Congress:75th IFLA General Conference & Assembly*. Torino, Milan, Italy.

28. Sannwald, W. W. (2001). *Checklist of library building design consideration* (4th Edition ed.). Chicago: American Library Association.

29. Shill, H. B., & Tonner, S. (2003). Creating a Better Place: Physical Improvements In Academic Libraries, 1995-2002. *College & Research Libraries, 64(6)*, 431-466.

30. Shill, H. B., & Tonner, S. (2004). Does the Building Still Matter?Usage patterns in new, expanded and renovated libraries,1995- 2002. *College & Research Libraries , 65(2)*, 123-150.

31. Yildirim, K., Akalin-Baskaya, A., & Hidayetoglu, M. (2007). Effects of indoor color on mood and cognitive performance. *Science Direct , 42* (9), 3233-3240.

Chapter 8

Social Navigation For Educational Digital Libraries

Peter Brusilovsky[a,*,] **Lillian N. Casselc, Lois M. L. Delcambreb, Edward A. Foxd, Richard**

[a] *University of Pittsburgh, PA 15217*

[b] *Portland State University, Portland, OR 97207*

[c] *Villanova University, Villanova, PA 19085*

[d] *Virginia Tech, Blacksburg, VA 24061*

[e] *Texas A&M University, College Station, TX 77843*

[f] *University of California, Berkeley, CA 94720-4600*

ABSTRACT

An educational digital library may store a wealth of diverse educational resources targeting different audiences from young schoolchildren to graduate students to school and college teachers. With the growth of the volume and the diversity of the library, it becomes increasingly difficult for the users to find resources, which are relevant to their age, educational needs, and personal interests. Social navigation techniques could provide valuable help in this context guiding users to the most useful information. Social navigation works by processing traces of past user behavior and using the assembled "collective wisdom" to guide future users. The paper reports our work on the design of social navigation infrastructure for Ensemble, a major educational digital library. We present the organization of both sides of the social navigation process: how social wisdom is collected and how it can be used to guide portal users to valuable resources. We also report the results of our most recent evaluation of the social navigation infrastructure.

Keywords: social navigation, digital library, portal, navigation support

INTRODUCTION

Recommender systems and other kinds of personalization is one of the recognized priority research direction in the field of Digital Libraries [4]. Personalization is especially important for educational digital libraries (EDL). The results of several research projects show that both, the ability to formulate good queries in the search process and the ability to choose appropriate links in the navigation process requires relatively high level of background or subject knowledge [3]. We can hardly expect that less prepared students will be able to locate relevant resources even in relatively small EDL where a query typically returns dozens of hits. However, it's exactly these less prepared students, including users coming from underrepresented groups, who need most the access to relevant high quality material. With further accelerated growth of digital libraries traditional ways of access may become challenging even for domain experts.

A traditional university library provides a number of ways to support students locating relevant resources. In addition to universal search interfaces that stay the same for every patron, libraries support several ways of personalization. Teachers use course reserves to provide their students with easy access to the most relevant books. Personal reserves allow students storing once found books so that next time they can start working with these books without search. Library administration can adapt to the needs of a specific group of patrons (e.g., Biology freshmen) by organizing focused sections that contain a high proportion of books relevant for this group. Traditional libraries also help students to find relevant books by social navigation [7]. Following their peers with similar needs and interests, students can easily locate sections and shelves where relevant books are located. "Wear and tear" of different books directs them to the most used resources. Marked book pages and pencil comments allow finding fragments that are important for a specific course. (For that reason many student are known to prefer used textbooks.) Finally, the highest level of personalized support can be provided by an expert reference librarian. A good librarian can recommend relevant books for a specific course that were overlooked by a teacher or that arrived recently. She can examine a pile of books selected by a student and suggest which ones will be most helpful to the student's course needs and interests. She can also recommend books that are similar by content to selected books or books that are typically picked out by other students with similar needs and interests.

It has been already recognize that EDL should attempt to replicate known personalization approaches to serve their users better. A number of recent projects have already started to explore various ways of to offer personalized experience to the users of such libraries [8; 11; 14; 16]. For our NSDL Ensemble project [9], which is a large collaborative effort to build an EDL on computing, personalized support was one of the main goals. Our first target is social navigation mentioned above. Social navigation works by processing traces of past user behavior and using the assembled "collective wisdom" to guide future users. There are two forms of social navigation: collaborative filtering and history-enriched information spaces [7]. While the majority of projects exploring social navigation are exploring collaborative filtering, we attempt to employ both groups of techniques.

The Ensemble *portal* is engineered to collect various traces of user interaction with the portal and to use this wisdom to guide future users of the portal using adaptive link annotation and link recommendation. This paper presents this process in detail. It explains how social wisdom is collected by various portal components and how it can be used to guide portal users. It also reports the results of our most recent evaluation of the social navigation infrastructure.

ENSEMBLE: A PORTAL FOR COMPUTING EDUCATION

Ensemble is the Computing Portal in the US National Science Digital Library (NSDL), an ambitious project to provide access to learning materials and resources for education in the Science, Technology, Engineering and Mathematics (STEM) disciplines at all age and education levels. Ensemble provides a distributed portal to such materials, allowing both multiple sources and also multiple access points to those resources. In these aspects, Ensemble is similar to other large-scale projects focusing on developing focused educational portals such as MACE [17] or Organic.Edunet [11]. Ensemble explicitly supports the role of a library as more than a repository by emphasizing a role as a gathering place, a place for sharing, for finding others with similar needs and interests, a place for meeting to work on a common project, a place for highlighting people and resources of particular interest.

The key component of Ensemble is the Web portal: www.computingportal.org. The portal presents easy access to recognized collections and to tools found useful by computing educators. It is also a central meeting place for communities who are interested in various aspects of computing education. The three themes of the portal: collections, communities, and tools, are interrelated and together represent a social connection among people and between people and resources. This social interconnection is the focus of this paper.

SOCIAL NAVIGATION IN ENSEMBLE: AN OVERVIEW

The collective wisdom in Ensemble is assembled by tracking various actions of portal users. The Ensemble portal tracks both traditional low-level user actions such as resource browsing, rating, commenting, and tagging; and higher level structural actions such as fragment extraction and composition. The information about all these actions is accumulated in the user modeling server of the portal, which processes this information and makes it available to social navigation services (Fig. 1).

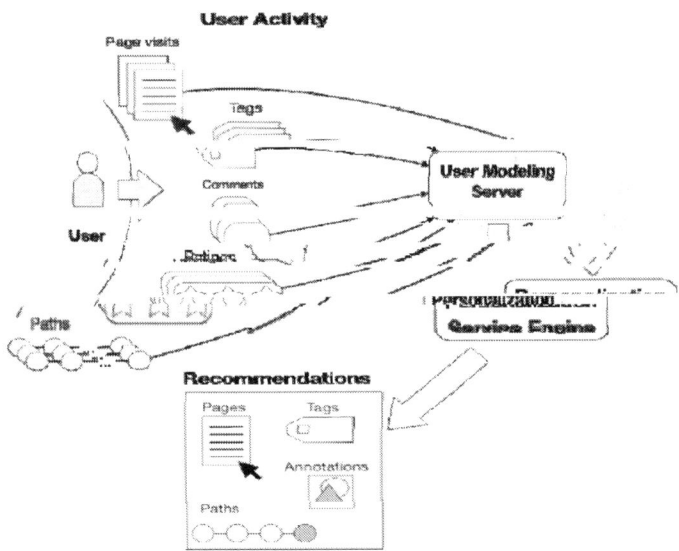

Figure 1. The social navigation architecture of Ensemble portal. Various traces of user actions are collected and processed by the user modeling server. The accumulated "wisdom" of the user community is used by personalization services to provide social navigation.

Following the nature of Ensemble as a meeting point for various groups interested in computing education, the Ensemble portal processes the collective wisdom on two levels: the *portal level* and the *group level*. The portal level integrates traces of all portal users, while the group level integrates actions of a specific community or group of users such as, for example, a CS1 community, a group of users focused on the first-year course in computer science (http://www.computingportal.org/cs1). Each user working with Ensemble can explore the portal as a member of such a group. In this case, each action of this user contributes to both the portal wisdom and the wisdom of her current group. In turn, an association with a specific group enables group-level social navigation, i.e., exploring the portal as a member of a group, the user is guided by both portal-level and group-level wisdom.

Group association is neither mandatory nor exclusive. A user can explore the portal without a group association (in this case only portal-level wisdom is used). A user can also belong to several groups and communities and change her current group at any time. However, at any given time the user can be dynamically associated with one group and explore the portal from one perspective since social navigation guidance offered by different groups could be contradictory.

COLLECTING SOCIAL WISDOM IN ENSEMBLE

Ensemble integrates experiences of several earlier social navigation projects and tracks a wider set of user actions than earlier systems. Feedback actions, which are traditionally used in social navigation systems, are focused on a specific resource and provide explicit or implicit feedback on that resource. These actions can be used to estimate user and group level of interest in the resource. Application actions comprise using resources in the context of user needs. The usage is traced on both the level above and below stored resources. The level above means composing artifacts from several resources (such as the Ensemble guided path mechanism). The level below means extracting a fragment of the resource for re-use. Tracking these actions allows Ensemble to support more sophisticated social navigation.

Tracking User Navigation

Ensemble uses Drupal, a content management system that has gained popularity for its dynamic features. Instead of traditional log files, Drupal uses database tables to store user activities. The tracking is done both on anonymous and authenticated users. Based on user-id, we can track what pages a user viewed on a particular date. It also lists the hostname, which can be used to track down the geographic origin of a page request. Tracking user browsing outside the Drupal portal is more complicated and can be done by using *embedlets* (see section 6) and Google Analytics. Both kinds of browsing history can be processed by the user modeling server and applied to provide social navigation. For example, link annotations attract user attention to resources that are most popular among the group users, while link generation can recommend resources that were visited next by users from the same group in a similar context. More details on link annotation and generation can be found in section 6.

User Direct Feedback: Comments, Ratings, Tagging

A modern state-of-the-art educational portal is expected to offer three kinds of user feedback, known as CRT (Comments, Ratings, Tagging) [1; 17]. Comments and ratings are examples of explicit feedback; they allow users who have enough interest in a resource or a post to augment the content in some way. These types of feedback originated from e-commerce sites such as Amazon.com, where ratings and reviews by visitors with experience with the product serve to identify characteristics of the product that will be of interest to potential buyers. Brought to the DL world by pioneer projects such as Merlot [12], these forms of feedback demonstrated their ability to guide portal users to most useful resources.

Comments are the most extensive. The user contributes something to the item by extending the information provided or by providing feedback to indicate the strengths or the weaknesses of what has been posted. The comment can be substantive to the point that it rates nearly as high in importance as the original submission, or the comment can be a simple reinforcement of what was submitted previously. Ratings are brief and require minimal effort on the part of the visitor. As implemented in the Ensemble portal, ratings mean indicating a score of 1 to 5 stars, with the number of stars corresponding to the quality of the item in the opinion of the visitor. A rating that does not include a comment may not be of great value for an end user. However, due to their numeric nature, ratings provide the most usable information for collaborative filtering and social navigation.

Tagging is the labeling of the item that allows the users to organize information. In the Ensemble project, we have provided both controlled vocabulary and free form (community) tagging. The advantage of the controlled vocabulary, here specifically the computing ontology [5], is that the same word or phrase will be used consistently, aiding in effective search and classification. The advantage of community tagging is the ability of the user to express what they believe are the relevant features of the item without having to compromise their expression when the controlled vocabulary does not match their need very well. Both are valuable and we will explore the usefulness of both as the project progresses.

In the social navigation component of Ensemble, the CRT feedback is used in two ways. First, they all indicate user interest in various items. Second, they allow the system to identify items that are similar from the user behavior perspective. This data is used to generate links to similar resources.

Fragment Selection: Superimposed Information in Ensemble

One of the main goals of Ensemble is to enhance the application of stored information by providing users access to content both below and above the stored level. Ensemble provides support for *superimposed information* - the use of fine-grained fragments from existing data sources in new contexts. The main concept is that potentially fine- grained selections from an existing document or digital object can be referenced (using an appropriate addressing scheme) and then used in a variety of ways. We and our colleagues have built a variety of tools that allow users to select portions of MS Office, XML, PDF, and HTML documents and structure, annotate, and elaborate them in a scratchpad tool [6]. For example, users can highlight and label appropriate selections from an electronic copy of a textbook with the appropriate learning objective from a university database course [2]. Our data shows that superimposed information is useful for end users and information access tools (i.e., to improve document [13]).

Ensemble implements support for superimposed information using Fedora. The implementation allows a user easily to create what we call subdocuments – the selected portion of an existing digital object – as a first-class object in a digital library. An author can select a portion of a web resource, annotate it graphically in situ, and then save the resulting annotated excerpt as an object in a digital library, along with metadata that specifies the address of the entire original resource, including a link to the specific section used. For each such object, we also create a view consisting of a web page that displays the annotated excerpt along with a link to the original resource. Our work enables all user actions against subdocuments to be logged. More than that, Ensemble can track the creation of subdocuments as well as their use in more complex, superimposed digital objects, including Waldens' Paths, described in the next section. The information about fragment extraction, annotation, and reuse provides valuable information for social navigation mechanisms of Ensemble.

Feedback by Composing: Waldens' Paths

In addition to the ability to extract information, Ensemble allows the user to compose information active *above* the storage level. This ability is supported by a guided path mechanism known as Walden's Paths [15]. A guided path is an organized and annotated collection of resources composed by an end user. Superficially a path has many attributes in common with a resource list, since it is essentially a collection of links with annotations like some resource lists. However, the path provides a collection of resources external to the path but still part of the path, unlike resource lists, the *in situ* nature of its annotation makes it possible for an author to form a narrative structure while still containing the resources of interest.

This system provides users with authoring and viewing interfaces. The authoring interface allows users to compile their resources and add annotations in order to construct a narrative structure while maintaining the original form of the individual resources. The resources and their annotations create a linear structure, the better to facilitate the path creator's constructed narrative. After the user has completed authoring a path, they may chose to make it publicly available or may choose to provide the path's URL directly to the other users. The viewing interface allows users to view paths. When viewing a path, the user sees the original resource inside of the interface as well as an external hyperlink containing the original resource's URL. If the user navigates off of the guided path the navigational interface changes to notify the user that he has ventured off the guided path. At any point, if the user wants to return to the guided path from his self-guided exploration, he may click the "back to path" button to return to the point in the path where he last left off.

In practice, a path acts as a communication medium through which a creator can interact with a user. A creator can be a teacher providing education materials to a class or to other teachers, or a student creating a path to demonstrate their understanding of a topic to the teacher or other students— indeed in our experiences with Walden's Paths we have observed all of these cases. Outside of the formal educational setting, a path author can be any person who is interested in providing a constructed narrative around a set of web resources. In essence, Walden's Paths becomes the intermediary through which previously disconnected resources are contextualized and communicated. Being important in DL context for a range of reasons, Walden's Paths serve as an invaluable source of information for social navigation. The creator's decision to include a resource into a guided path indicates user and group interest in this resource, which can be used for social annotation. The co-location of items in various paths and their annotation provide a reliable way to estimate resource similarity from a perspective of a specific group. The discovery of this similarity is critical to generating links to recommended items.

THE MECHANISMS FOR SOCIAL NAVIGATION

To account for the distributed nature of Ensemble, the portal uses a service-based personalization approach. All kinds of social navigation are provided by the personalization engine PERSEUS, which is independent from the portal. As a result, social navigation to portal resources can be provided not only inside the main portal, but also within its various components and tools (such as Walden's Paths) and even on personal sites of users who apply portal resources for their needs.

The PERSEUS engine is able to offer social navigation support for any list of links to resources inside or outside the portal. Inside the portal, these lists of links can be found in the terminal nodes (leaves) of the portal's traditional hierarchical organization. In this case, this list of links corresponds to a group of similar resources collected and classified by the portal, for example, a group of links to applets demonstrating various sorting algorithms. Outside the portal, this list of links can appear, for example, on a Web site of the instructor as resources assembled to support a specific lecture.

To extend this list of resources with social navigation support, the list should be included in an *embedlet*, a specific fragment of HTML, which is responsible for both tracking user browsing and generating social guidance. When a Web browser renders this embedlet, the list of resources along with the information about the current user and her group is sent to PERSEUS. PERSEUS contacts the user modeling server and produces two kinds of social navigation support for this list: social visual cues and personalized recommendations (Fig. 2). *Social visual cues* annotate links in the original list showing their relevance to the current community. PERSEUS can provide several kinds of visual cues, however so far, ENSEMBLE uses only one kind, the traditional "wear" cues [10] that indicates the resources that are most popular in a group by color intensity (Fig. 3). The more intense the color is, the more popular is the resource. *Personalized recommendation* extends the original list with other resources that are considered similar to the resources in this list from the perspective of the current group. To distinguish between the original and the recommended resources, the links to recommended resources are annotated by the star icon (Fig. 3).

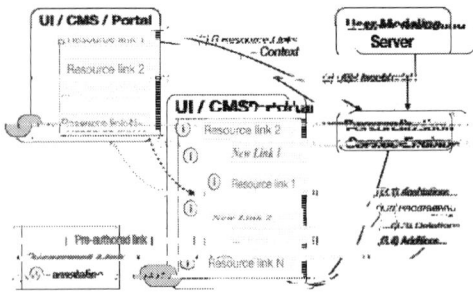

Figure. 2. Scenario of personalization service engine's operation.

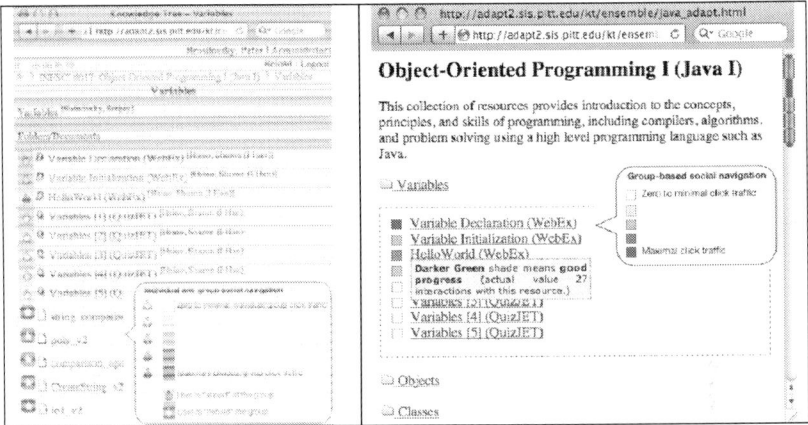

Figure 3. Social navigation support on a portal (left) and a static web page (right).

THE EVALUATION OF THE SOCIAL NAVIGATION ARCHITECTURE

To test the feasibility of the service-based social navigation approach developed for the Ensemble portal, we performed an experimental evaluation of PERSEUS working in a social navigation service mode. The goal of this evaluation was to test PERSEUS under heavy loads and determine its performance limits. To accomplish this, we used a technique discussed in [19]. We set up two machines: one with the PERSEUS engine, and the other with a special "flooder". The role of the "flooder" was to imitate the personalization service client, subject the personalization engine machine to various types of load, and record the observed parameters.

To simplify the experiment, discrete values of load parameters were used. The parameters of the load were the following.

- **Complexity of the request**. The majority of our portal lecture folders had roughly 20 resource links to personalize. In addition, we used two more values for complexity: 5 resources, to represent a "lightweight" folder, and 50 to signify a folder "overloaded" with resources.

- **Request delivery rate** – delay between consecutive requests. From our experience with user modeling services [19] and initial experiments with a personalization service engine, we had already learned that a delay of 10ms between requests is critical for our hardware/software configuration. In addition, delays between requests of 160ms and more did not present any challenge. Hence, we varied the request delivery rate parameter between 10 ms and 160 ms. Rates between boundaries were doubles of the previous value, giving us 5 loads: 10, 20, 40, 80, and 160 ms.

- **Duration of the load**. From prior experimentation we knew that the duration did not really matter, unless it was a peak load of 10ms or 20ms between requests. During these peak loads, the personalization server would stop responding to any requests at all after 30 seconds. We decided to keep the load sessions fairly short – a little less than 4 seconds (3,840ms, divisible by all delivery rates).

To obtain more data we repeated the flooding sessions 5 times for each of the three request complexities and each of the five request delivery rates, giving us 3 x 5 = 15 different settings. During these sessions we observed the following parameters:

- **Mean response delay**—the average amount of time it takes to complete a request.

- **Request success rate**—the fraction of requests that completed successfully. For the least demanding load of 160 ms between requests, the amount of requests sent per each flooding session was 3,840/160=24. For the highest load of 10 ms between requests, it was 3,840/10=384.

The personalization service engine was run on a Pentium 4 dual core 2.8 Mhz processor with 1Gb RAM. The user modeling server that the personalization service engine depended on was running on the same machine. To compensate for the high speed of wired intranet, we used a WiFi network to communicate with the personalization engine. It also provided a realistic scenario for students who would be accessing adaptive content outside their fast university campus LAN.

In our experiments we were trying to answer three questions. First, what is the highest load the PERSEUS's social navigation service can cope with without performance degradation. Second, how large is the user population that it can effectively support. And third, what is the share of PERSEUS in the overall request delay and when is this share significantly different from network-induced delay.

The answer to the first question is summarized in Table 1. Depending on the request complexity the highest acceptable loads were 20, 40, and 80 ms between requests to provide social navigation for 5, 20, and 50 resources, respectively. In all three cases, PERSEUS's response was quite fast. In the worst case (50 resources per request) 95% of the requests finished in 200ms, which is nearly instantaneous for a human observer.

Table 1. Summary of performance evaluation results

Request complexity	Highest acceptable load	Time to complete 95% of requests
5 resources	20 ms between requests	25 ms
20 resources	40 ms between requests	50 ms
50 resources	80 ms between requests	200 ms

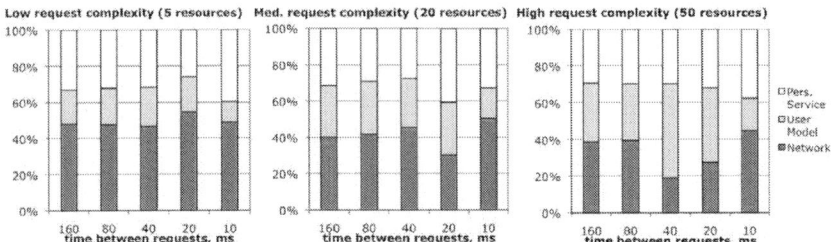

Figure 4. Request time distribution: network communication, CUMULATE and PERSEUS.

While attempting to answer the second question – size of the supported user population - we based our estimation on the user activity observed in our earlier classroom studies [18]. A class of 37 students working together 95% of the time produced fewer than 7 requests per minute. Since a typical page request contained roughly 20 resources that needed social annotation, we based the estimation on the results for the medium complexity requests obtained above. A load of 40 ms between requests was previously confirmed to be the maximum PERSEUS can effectively handle in this context. 40ms between requests is 60,000 / 40 = 1,500 per minute. Our class of 37 students produced a maximum 7 requests per minute, whenever they were working with the system. Thus, to reach the allowed request maximum of 1,500, (37*1500 / 7) ≈ 8,000 students have to be actively involved.

The third and last question pertaining to the PERSEUS's performance is its timeshare in the overall page- rendering delay. We were comparing PERSEUS's timeshare to the cumulative network delay and the timeshare of the user modeling server CUMULATE [19] used by PERSEUS. As it turned out, the split between the three was roughly equal, without significant advantage of one or the other across all request complexities and delivery rates.

SUMMARY AND FUTURE WORK

This paper argued for the need to extend digital libraries with social navigation features and demonstrated how social navigation can be implemented in the context of a large distributed educational DL, the Ensemble computing portal. We demonstrated a range of user interactions, which can be tracked to form the collective wisdom and presented an architecture that is able to apply this wisdom to guide future users of Ensemble. The social navigation approach is already used in some components of the portal. The evaluation shows that even a single-server implementation can support much a larger volume of users than we have now.

In the future, we plan to extend a range of social information access approaches offered by Ensemble by adding collaborative filtering and social search.

ACKNOWLEDGEMENTS

This material is based upon work supported by the National Science Foundation under Grant Numbers 0534762, DUE-0840713, 0840719, 0840721, 0840668, 0840597, 0840715, 0511050, 0836940 and 0937863.

REFERENCES

1. Abel, F., Marenzi, I., Nejdl, W., Zerr, S.: Sharing Distributed Resources in LearnWeb2.0. In: Cress, U., Dimitrova, V., Specht, M. (eds.) Proc. of 4th European Conference on Technology Enhanced Learning (ECTEL 2009). Lecture Notes in Computer Science, Vol. 5794. Springer-Verlag (2009) 154-159

2. Archer, D.W., Delcambre, L.M.L., Corubolo, F., Cassel, L.N., Price, S., Murthy, U., Maier, D., Fox, E.A., Murthy, S., McCall, J., Kuchibhotla, K., Suryavanshi, R.: Superimposed Information Architecture for Digital Libraries. In: Proc. of 12th European conference on Research and Advanced Technology for Digital Libraries (ECDL'08). Lecture Notes in

Computer Science, Vol. 5173. Springer-Verlag (2008) 88-99

3. Brusilovsky, P.: Adaptive navigation support in educational hypermedia: the role of student knowledge level and the case for meta- adaptation. British Journal of Educational Technology 34, 4 (2003) 487-497

4. Callan, J., Smeaton, A., Beaulieu, M., Borlund, P., Brusilovsky, P., Chalmers, M., Lynch, C., Riedl, J., Smyth, B.: Personalisation and Recommender Systems in Digital Libraries, Joint NSF-EU DELOS Working Group Report (2003)

5. Cassel, L.N., Davies, G., Fone, W., Hacquebard, A., Impagliazzo, J., LeBlanc, R., Little, J.C., McGettrick, A., Pedrona, M.: The computing ontology: application in education In: Working group reports of Innovation and technology in computer science education, (ITiCSE). Dundee, Scotland (2007) 171-183

6. Delcambre, L.M.L., Maier, D., Bowers, S., Weaver, M., Longxing, D., Gorman, P., Ash, J., Lavelle, M., Lyman, J.: Bundles in Captivity: An Application of Superimposed Information. In: Proc. of 17th International Conference on Data Engineering. Lecture Notes in Computer Science, Vol. 5173. Springer-Verlag (2001) 88-99

7. Dieberger, A., Dourish, P., Höök, K., Resnick, P., Wexelblat, A.: Social navigation: Techniques for building more usable systems. interactions 7, 6 (2000) 36-45

8. Drachsler, H., Pecceu, D., Arts, T., Hutten, E., Rutledge, L., Rosmalen, P.v., Hummel, H., Koper, R.: ReMashed – Recommendations for Mash-Up Personal Learning Environments. In: Cress, U., Dimitrova, V., Specht, M. (eds.) Proc. of 4th European Conference on Technology Enhanced Learning (ECTEL 2009). Lecture Notes in Computer Science, Vol. 5794. Springer-Verlag (2009) 788-793

9. Fox, E.A., Chen, Y., Akbar, M., Shaffer, C.A., Edwards, S.H., Delcambre, L., Decker, F., Archer, D., Brusilovsky, P., Furuta, R.,

Shipman, F., Carpenter, S., Garcia, D., Cassel, L.: Ensemble PDP-8: Eight Principles for Distributed Portals. In: Proc. of The 10th ACM/IEEE-CS Joint Conference on Digital Libraries. ACM Press (2010) 341-344

10. Hill, W.C., Hollan, J.D., Wroblewski, D., McCandless, T.: Edit wear and read wear. In: Proc. of SIGCHI Conference on Human Factors in Computing Systems, CHI'92. ACM Press (1992) 3-9

11. Manouselis, N., Kosmopoulos, T., Kastrantas, K.: Developing a Recommendation Web Service for a Federation of Learning Repositories. In: Proc. of 2009 International Conference on Intelligent Networking and Collaborative Systems. IEEE (2009) 208-211

12. Nesbit, J., Belfer, K., Vargo, J.: A Convergent Participation Model for Evaluation of Learning Objects. Canadian Journal of Learning and Technology 28, 3 (2002)

13. Price, S., Nielsen, M., Delcambre, L., Vedsted, P., Steinhauer, J.: Using semantic components to search for domain-specific documents: An evaluation from the system perspective and the user perspective. Information Systems 34, 8 (2009) 724-752

14. Ruiz-Iniesta, A., Jiménez-Díaz, G., Gómez-Albarrán, M.: User-Adaptive Recommendation Techniques in Repositories of Learning Objects: Combining Long-Term and Short-Term Learning Goals. In: Cress, U., Dimitrova, V., Specht, M. (eds.) Proc. of 4th European Conference on Technology Enhanced Learning (ECTEL 2009). Lecture Notes in Computer Science, Vol. 5794. Springer-Verlag (2009) 645-650

15. Shipman, F., Furuta, R., Brenner, D., Chung, C., H., H.: Guided Paths through Web-Based Collections: Design, Experiences, and Adaptations. Journal of the American Society for Information Science 51, 3 (2000) 260-272

16. Verpoorten, D., Glahn, C., Kravcik, M., Ternier, S., Specht, M.: Personalisation of Learning in Virtual Learning

Environments. In: Cress, U., Dimitrova, V., Specht, M. (eds.) Proc. of 4th European Conference on Technology Enhanced Learning (ECTEL 2009). Lecture Notes in Computer Science, Vol. 5794. Springer-Verlag (2009) 52-66

17. Wolpers, M., Memmel, M., Giretti, A.: Metadata in Architecture Education - First Evaluation Results of the MACE System. In: Cress, U., Dimitrova, V., Specht, M. (eds.) Proc. of 4th European Conference on Technology Enhanced Learning (ECTEL 2009). Lecture Notes in Computer Science, Vol. 5794. Springer-Verlag (2009) 112-126

18. Yudelson, M., Brusilovsky, P.: Adaptive Link Annotation in Distributed Hypermedia Systems: The Evaluation of a Service-Based Approach. In: Nejdl, W., Kay, J., Pu, P., Herder, E. (eds.) Proc. of 5th International Conference on Adaptive Hypermedia and Adaptive Web-Based Systems (AH'2008). Lecture Notes in Computer Science, Vol. 5149. Springer Verlag (2008) 245-254

19. Yudelson, M., Brusilovsky, P., Zadorozhny, V.: A User Modeling Server for Contemporary Adaptive Hypermedia: An Evaluation of Push Approach to Evidence Propagation. In: Conati, C., McCoy, K., Paliouras, G. (eds.) Proc. of 11th International Conference on User Modeling, UM 2007. Lecture Notes in Computer Science, Vol. 4511. Springer Verlag (2007) 27-36

Chapter 9

Qualitative Assessment of Mould Growth for Higher Education Library Building in Malaysia

Suriani Ngah Abdul Wahab[a*], Nurul Izma Mohammed[a], Mohd Faris Khamidi[b], Nazhatulzalkis Jamaluddin[a]

[a] Faculty of Engineering, Universiti Teknologi PETRONAS, Bandar Seri Iskandar, 31750 Tronoh, Perak, Malaysia

[b] School of the Built Environment, Heriot-Watt University Malaysia, Level 2, Menara PjH, 2 Jalan Tun Abdul Razak, Precinct 2, 62100 Putrajaya, Malaysia

ABSTRACT

A qualitative assessment that identifies factors to support the mould growth in the library building and makes a recommendation for correcting these factors provide helpful guidance for the librarian and the library users. The aim of this paper is to examine whether qualitative assessment of mould growth can provide better information about library environments. The information gathered from the walk-through, interviews, and ventilation review using Microclimate data logger of three university libraries in Malaysia. Researcher was conducting a visual inspection of the immediate building interior and exterior including ventilation and air conditioning system. In conclusion, applicable qualitative assessment of mould can be use and sufficient to begin planning appropriate improvements to the environment.

Keywords: mould growth; library; qualitative assessment; visual inspecti

INTRODUCTION

Academic library is an important space at the University, and excellent library space is one of the significant aspects that need to be measured in promoting library utilization (Atmodiwirjo & Yatmo, 2011). Previous research has shown it has become apparent that moisture and humidity control problem exists in library buildings. The cause of these problems can be complex and involve many aspects of library design, construction and maintenance. Improper humidity control inside a library can lead to material deterioration and the development of mould. According to (Chen & Garcia, 2004) many of the problems leading to mould and inadequate humidity control can easily be identified and prevented. The study identified humidity problem and classified it into four main humidity problems such as design, construction, retrofit and alterations and poor maintenance. Once the mould has grown, it can be easily identified with the human eye. Therefore, in a qualitative assessment the researcher will evaluate the information gathered from the walkthrough, interviews and ventilation review. The researcher seeks to identify sources of mould growth and to define the pathways in the environment that may bring mould and any associated toxins into contact with the building occupants (Karbowska-Berent, Górny, Strzelczyk, & Wlazło, 2011; Rao, Burge, & Chang, 1996; Wang, Chen, & Zhang, 2010). Mould can survive and grow when indoor relative humidity is high and fungus is present all the time in the air that we breathe. According to (Portnoy, Barnes, & Kennedy, 2004), Three reasons we need to worry about expose to the mould in the indoor environment. The first is possible health effects to its exposure and also their metabolic products, second the effect on the material and structural such as deterioration of the material and the final is the depressing aesthetic effects. The spores can be easily inhaled, which can cause the following symptoms: coughing, wheezing, runny nose/sinus problems, skin rashes if touched, and other immune responses after long-term exposure. People who experienced chronic above symptom could risk to asthma. However, the type of severity of health symptoms depends upon the type of mould, the amount of exposure, ages and our fitness(Illinois Department of Public Health, n.d.). Children asthma is foremost clinical fear worldwide (Al, Ali, & Thamiri, 2011)

Humidity control has been a major concern in the indoor library environment. Excess humidity can allow for the growth of the mould in these library buildings. Libraries have been storing and collecting depiction obtains more thousands of books and documents. The situation of such conserved books collection is under the stable control of different environmental factors. The risk of this situation is the climate in the library itself such as temperature and humidity of the air, type and amount of light received by collections. A qualitative assessment of the ventilation system is a key aspect of assessing the library user's environments. Ducts with internal lining or duct board can become microbial reservoirs and amplifiers if they become humid and dirty (Storey et al., 2004). The presence of mould growth using qualitative assessment was the primary focus of this study, in which the researcher presents the evidence from the three different cases studies.

LITERATURE REVIEW

Biology Of Mould and Sampling Analysis

Mould or Microscopic organisms found virtually everywhere, indoors and outdoors. Fungal growth requires oxygen, adequate temperature, nutrients and water. As they grow, they reproduce by producing tiny spores that float through the air, spores are widespread in the environment and will drift indoors. If these spores land on wet or moist organic material, they will grow, producing fungal colonies that decay the organic matter to stay alive. This situation can cause some property damage and present health problems.

Various authority and agency in their guidelines in assessing mould growth not recommended to do sampling due to several reason such as a) present visible mould would enough for remediation strategies; b) testing and analysis of mould is expensive; and c) assessor should be have a clear reason for doing so (Environmental Protection Agency (EPA), 2010). However, sampling may help the assessor to find the source of mould contamination, identify present species and differentiate between mould and dirt. American Hygiene Association (AIHA), 2008 also suggested that mould sampling should be performed only when a visual inspection reveals issues or the presence of mould cannot be confirmed by visual inspection. The reason of sampling is to attain an approximate of the concentration of viable mould spores or mould material per unit area of surface or per unit of volume air. One familiar type of air sampler passes a measured quantity of air through a petri dish, which is sent to a laboratory for culture. The cultured spores allocate individual species to be determined, and the results are reported out as colony-forming units (CFUs) per cubic meter. It grows in different species in different agars. Therefore, selection of agar can influence the result regarding the species reported (B.Rose, 2005). However newly, PCR methods (DNA sampling) have been used for very precise and accurate estimates of a quantity as well as species of mould material in samples.

Mould Sampling Method

The sampling method chosen depends on resource availability, method, availability and information required (AIHA, 2008). In general, mould sampling method used to collect mould from air, surfaces and settles dust. The summarized of Mould Sampling Method is shown in Table 1 below.

Table 1. The Summarized of Mould Sampling Method

No	Method	Description / Types	Unit of Measurement	Strength	Weakness	References
1	Air sampling method	i) *Bioaerosol air sampling.* Identify type of fungal species present, number of fungal colonies and percentage of each type of species in total of all microbial species found. *ii) Spore trap air sampling.* Identify total number of viable living and non-viable fungal structure or	Cfu/m³ (Colony forming unit) S/m³ (Struct ure) C/m³	Quick, easy sampling method. May indicate mould growth present that is no visible.	Samples from dusty areas may be overload with particles.	NYCOSH Carolina Environment al, Inc
		colonies per cubic meter of air sampled. Indoor type and concentrations should reflect outdoor type and concentration. iii) Micro-vac sampling	(Colony) Spores/g Cfu/g			
2	Surface sampling method	Used direct macroscopic examination. i) Surface wipe sampling ii) Surface swab sampling iii)Tape lift samplings Applies to the optical analysis of surface samples for vegetative and reproductive structures/fruiting bodies of fungi.	Cfu/gram Cfu/inch² Cfu/cm² Percentag e (%)	Quick, easy sampling method. Identifies viable moulds that are present. Tape lift samplings allow accurate characterization of which spores. Reveal spores that have not yet become airborne.	Not effective on porous surfaces. Quantification can be significantly affected by sampling technique.	NYCOSH Environment al Health Laboratory-ESIS,Inc. Carolina Environment al, Inc
3	Settle Dust	Reservoir dust from carpet or mattresses is collected. i) Dustfall collector	Weight/g of sample dust Weight/m²	Presume integration over time that occurs in deposition of the pollutant on surface. Cheap to produce and simple to use and microbial level measured with this	Surface samples allow only a crude measure that is probably only a poor surrogate for airborne	Institute of Medicine 2000

MATERIAL AND METHODS

Survey Procedure

The study was carried out in three different higher education institution libraries in Malaysia. Field work was carried out in November 2013 to March 2014. The sites were surveyed with the following criteria: collection of library building characterization data based on direct inspection and interviews with library employees; library characterization including measurement of temperature and humidity, lighting and air conditioning systems and physical examination of the building; and inspection of sites that has been wetted or showed mould growth. The researcher gathered qualitative data by interviewing the library personnel and maintenance personnel and taking a walk-through library tour. The walk-through will explore the immediate inside environments and the physical structure of the library building; note water or moisture symptoms from past and present leaks, spills and condensation; review ventilation and note apparent mould, and area with mouldy and musty odours. Likely places and library materials (books and archives) where moisture may accumulate, such as crawlspaces should be noted.

Case Study profile

Table 2 gives some insight into the case study profile in which the researcher has performed field studies and provided the qualitative assessment.

Table 2. Case study profile

Library Data	Case Study		
	Library TP	Library SA	Library TH
Year Built	2004	1976	1986
Built Up Area	12,790 sq. m.	NA	18,639.5 m sq
Location	Northern Malaysia	Central Malaysia	Southern Malaysia
Storey Height/Level	Ground Floor – Circulation Counter, Reference Collection, Malaysia Collection, Online Reference, Children Collection, Carrel, and Discussion Rooms Level 1 – Open Collection, Journal Collection, PG Lounge, IT Zone, Carrel, and Discussion Rooms Level 1M – Open Collection Level 2 – Open Collection, Office, IT Zone, Carrel, and Discussion Rooms Level 2M – Open Collection Level 3 – Open Collection, Multimedia Collection, Office, IT Training Zone, and Carrel	Level 1 – Office and Prayer Room Level 2 – Office, Meeting Room, and store Level 3 – Circulation Counter, Customer Service area, Office, and IT Zone Level 4 – Reading area, Open Collection, Customer Sevice room, and Reference Collection Level 5 – Reading area, Open Collection, and office	Level 2 – Reading area, Open Collection, Magazine collection, Discussion Rooms, Conference Room, and Reference Collection Level 3 – Circulation Counter, Reference Collection, Customer Service area, Office, Carrel Rooms, Meeting Room, Visual Audio Room, and IT Zone Level 4 – Reading area, Open Collection, Conference room, Carrel Rooms and Reference Collection Level 5 – Reading area, Open Collection, Reference Collection and Carrel Rooms
Books and library collection	213,067 vol	158,720 vol	534,100 vol

RESULTS AND FINDINGS

In the study, an attempt was made to determine the mould growth in a comprehensive manner by using qualitative assessment methods. The objective is to find areas where mould is growing and the possible effect of the dissemination into the breathing space in the library.

Visible mould growth

The qualitative assessment concerning visible mould growth and conditions to mould growth were assessed by the researcher during the field studies. One of the main causes of mould growth within the library buildings is poor maintenance. The mould growth was found at the upper level of the library building for Library SA and TH due to water damage in example water leakage from the roof. In Library SA, it was found evidence of water damage on the carpet due to failure of rainwater down pipe (rwdp) at Level 5 and it has located near to the bathroom at Level 4. The water penetrates from the outside wall also were recorded at Level 2, Library TH. As a result, the visible mould growth clearly appeared in this area. Table 3 lists a focused qualitative assessment of the immediate outside environment and building exterior for three libraries whilst Fig.1 shows the evidence of the described area.

Table 3. Focused qualitative assessment

Focused Qualitative Assessment	Visible Mould Growth		
	Library TP	Library SA	Library TH
Visual Assessment			
Source of outsides mould	X	√	√
Damage to the building	X	√ (water penetrate from the roof)	√ (water penetrate from the roof, wall)
Accumulations of organic material in or near air intakes	X	√	√
Occupied space survey			
Water damage	√	√ (roof leaks, rwdp)	√ (roof leaks)
Chronic condensation	X	X	X
Carpet	X	√ (carpet near toilets and cleaning activities)	X
Plants	x	x	√ (plant growth on flat roof)
Books and library collections	x	√ (dirt and dust, and microbial growth)	√ (dirt and dust, and microbial growth, cover damage)

Based on visual inspection it was found that some of books on the book rack located on the ground floor are identifying with dust on the book cover. The books were also swelling and shrinking. The researcher also found dirt, dust and microbial growth on the book that been displayed at the bookshelves in Library SA and TH as shown in Fig. 2.

Figure 1. Visible mould growth in the Libraries

Figure 2. Possible mould and microbial growth on the book surfaces.

Ventilation System Review

A systematic review of the mechanical ventilation system is part of the walk-through assessment. The way air moves in the library and the condition of the ventilating and air-conditioning system are critical aspects of bioaerosol exposure (Storey et al., 2004). Mechanical ventilation system should be properly maintained to optimize the volume of dilution air and to minimize the accumulation of contaminants that are microbial growth(Wang et al., 2010). In this review, the researcher had been assisted by the building's maintenance or mechanical engineering personnel. The personnel provide access to the ventilation system unit, provided maintenance history and describe the system's design parameters. Table 4 shows the review of the ventilated system used in the libraries and its condition.

Table 4. Ventilation system review

Ventilated System	Visible Mould Growth		
	Library TP	*Library SA*	*Library TH*
Operable windows	√ (available but for visioning purpose, closed all the time)	√ (available but for visioning purpose, closed all the time)	√ (available but for visioning purpose, closed all the time)
Air conditioning system	Centralised system	Centralised system	Centralised system
Maximum operational hours	8.30am – 10.00pm	24 hours	8.30am – 12.00am
Availability of Schematic drawings	√	√	√
Condition of Cooling Tower	√	X (mould visibly appeared)	X
Filters	√ (dampness, microbial growth, and dirt)	√ (dampness, microbial growth, and dirt)	√ (dampness, microbial growth, and dirt)
Air Diffusers	√ (dampness, microbial growth, and dirt)	√ (dampness, microbial growth, dirt and rust)	√ (dampness, microbial growth, dirt and rust)
Preventive Maintenance program	√	√	√
Air Purifier System	x	√	x
Portable Humidifiers	x	√	X
Exhaust Fan	√	x	X Available but not in use

Figure 3 shows the photo that were recorded during field studies regarding possibility of mould growth due to ventilation system

Figure 3. Sign of dampness, microbial growth and dirt

Measurement of temperature, humidity and lighting

Fig. 4 shows the measurement of temperature, humidity and lighting at ground floor, Library TP, which been recorded for four days. Based on fig. 4, the indoor air temperatures vary from 19.7ɛC to 24.8ɛC. The measured temperature at the book rack on ground floor varies from 19.7ɛC to 23.2ɛC, and is about 1ɛC to 2ɛC lower than the temperature measured at the book rack on the level 1, level 2 and level 3. Measured temperature at the book rack on level 1, level 2 and level 3 are equally similar with temperature varies from 21ɛC to 24.8ɛC. The average temperature for the ground floor is recorded at 21ɛC and becomes the lowest average temperature compared to other level in the Library TP building.

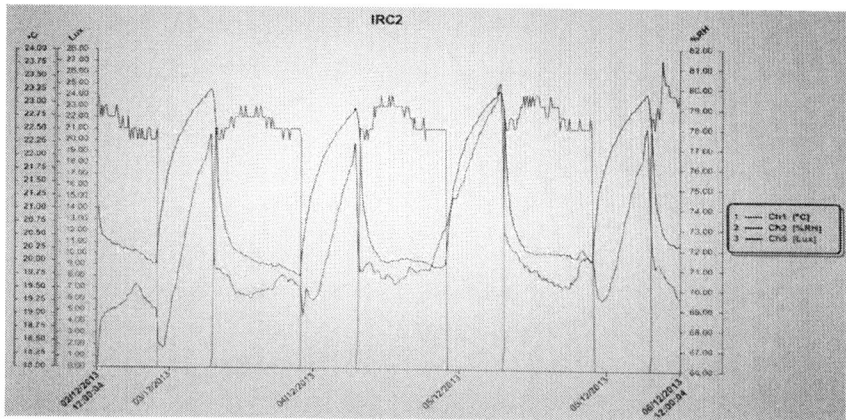

Figure 4. Data Recorded at Ground Floor Library TP

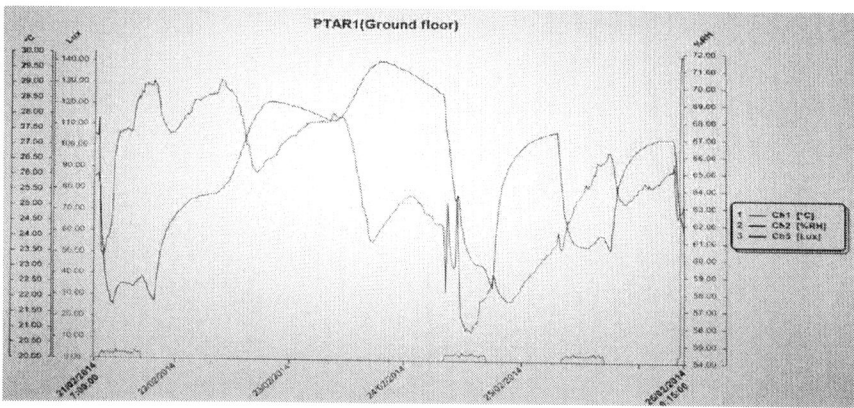

Figure 5. Data Recorded at Level 5 Library SA

The relative humidity in the library varies from 58.7% to 80.3%. The ground floor has been recorded as the place that has a high relative humidity with an average of 71.7%, and the maximum RH recorded early in the morning as high as 80.3%. The relative humidity increases consistently in the early morning from 6.30 am to 7.15 am. Overall relative humidity has been recorded at every level where it reached more than 60%. Mould can survive and grow when relative humidity reaches 60%, and germination requires higher relative humidity as well as time (Viitanen, 2007). Measured lighting varies from 0 to 79 lux.

Average lux was recorded from 259 to 219 lux. Fig. 5 shows the measurement of temperature, humidity and lighting at level 5, Library SA that been recorded for four days.

Based on fig. 5, the measured temperature at the book rack on level 5 varies from 21.8cC to 29.7cC. The average temperature for the level 5 is recorded at 26.3cC and becomes the highest average temperature compared to Library TP and TH. The relative humidity in level 5 has been recorded with an average of 64.3%, and the minimum RH recorded of 55.7% and the maximum RH recorded at 9.30 a.m. on 22nd February 2014 as high as 70.4%. Measured lighting varies from 0 to 143 lux and average lux was recorded at 149 lux. Fig. 6 shows the measurement of temperature, humidity and lighting at level 5, Library TH that been recorded for six days. Based on figure 3, the measured temperature at the book rack on level 5 varies from 23.2cC to 26cC. The average temperature for the level 5 is recorded at 24.7cC. The relative humidity in level 5 has been recorded with an average of 74.4%, and the minimum RH recorded of 66.6% and the maximum RH recorded at 8.00 a.m. on 29th March 2014 as high as 82.8%. Measured lighting varies from 0 to 80 lux and average lux was recorded from 31 lux.

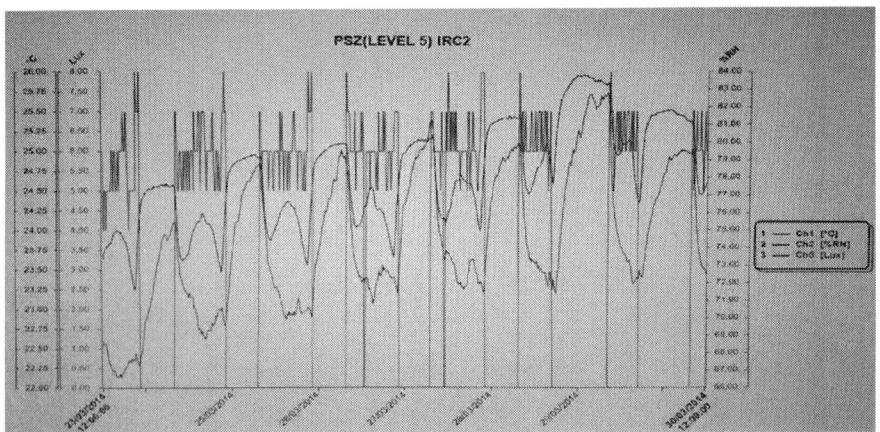

Figure 6. Data Recorded at Level 5 Library TH

DISCUSSION

Visible fungi and mould growth were higher in the upper levels of the Library SA and TH due to the ventilation problem and water damage. The library users and maintenance personnel reported mould, mouldy odors, wet or damp spots and water damage was higher at these particular areas. This phenomenon was extensively seen in the Level 5 at Library SA and TH. Built in the 1970's and 1980's, construction of roof and materials faced deterioration issues and poor maintenance. This evidence was a major factor for the growth of mould in the library building. In large institutions such as University SA and TH where there is a large number of buildings to maintain, regular scheduled maintenance is often delayed. Cost has become a major issue in any upgrading and maintenance work. As a result, problems that could have been prevented begin to accumulate and created deferred maintenance. In the example of Library TH, the main problems for the building were with the roof maintenance. During the walkthrough, the researcher found that the roof was not maintained properly, and unwanted plant and debris not been cleared by the maintenance contractor. As a result, the rainwater outlet and drainage was blockage. The water penetrates through the wall. However, it contrasts to Library TP that was built in 2004 where visible fungi and mould growth appear on the ground floor. It has been suspected that the affected area related to the high humidity that been recorded during the field study. The internal causes of the deterioration of papers in books are due to its acidity and biological factors (Pinzari, 2004; Zyska, 1997). The deteriorated books were found in Library SA and TH. Whereas not happen to Library TP. This situation relates to what has been told by Chief Librarian TP where all the books been fumigated before entering for the first time to the Library TP. In practice, fumigation has become the regular cure for books assumed to control insects or fungi or books in which an insect or fungus has been observed (Jan, 2006; Smith, n.d.) but not recommended because of human health concerns (Weaver-meyers, Stolt, & Kowaleski, 1998). Overall observation, the books that located at the crawls spaces higher risk of damage. It happens because of the lack ventilation of crawl spaces area. It was also shown at Library SA and TH, where the average temperature at the bookshelves is more than 24cC. However the lower the temperature, the better and its recommended between 18cC-22cC to preserved the books (Florian, 1994).

CONCLUSION

The development of mould growth in the library building came from many different sources. They are related with lack of humidity control, poor maintenance and related to building life itself. The existence of mould in the library building is a serious concern and should be treated immediately. Once the mould has been identified in the library, it should be avoided, and the cause to the growth should be identified and treated. Relying on the qualitative assessment can determine the visible of mould growth and causes to its development in the library buildings. However, the identification of mould species and types cannot be determined by using this method alone. Therefore in the next stage of the assessment, the author will take samples and identification of mould species and determines the possible solution to the mould growth in library buildings. Future research should concentrate on developing accurate, objective measures of exposure to mould growth.

ACKNOWLEDGEMENTS

Ministry Of Education and University Teknologi Mara sponsored the corresponding author and this paper. We greatly appreciate the assistance and useful discussion support from the Chief Librarians and Maintenance Personnel of Library TP, SA and TH.

REFERENCES

1.   Al, W., Ali, S. H., & Thamiri, D. Al. (2011). Pediatric Asthma and its Relation to SocioDemographic Factors in Baghdad. *Asian Journal of Environment-Behaviour Studies, 2,* 47–56.

2. Atmodiwirjo, P., & Yatmo, Y. A. (2011). Children's Participation in Library Space. *Asian Journal of Environment-Behaviour Studies*, 21–32.

3. B.Rose, W. (2005). *Water In Building - An Architect's Guide To Moisture and Mold* (p. 270). New Jersey: John Wiley & Sons, Inc. Chen, H., & Garcia, J. (2004). Roots of Mold Problems and Humidity Control Measures in Institutional Buildings with Pre-Existing Mold Condition.

4. Environmental Protection Agency (EPA). (2010). *A Brief Guide to Mold, Moisture, And Your Home* (p. 16). Washington D.C. Retrieved from http://www.epa.gov/mold/

5. Florian, M.-L. E. (1994). Conidial fungi (mould, mildew) biology: a basis for logical prevention, eradication and treatment for museum and archival collections. *Leather Conservation News*, *1*, 1–29.

6. Illinois Department of Public Health. (n.d.). Environmental Health Fact Sheet. Retrieved January 20, 2014, from http://www.idph.state.il.us/envhealth/factsheets/mold.htm

7. Jan, W. K. (2006). *Aerobiological Engineering Handbook A Guide to Airborne Disease and Control Technologies* (p. 846). McGRAW-HILL New York: McGraw-Hill Componies.

8. Karbowska-Berent, J., Górny, R. L., Strzelczyk, A. B., & Wlazło, A. (2011). Airborne and dust borne microorganisms in selected Polish libraries and archives. *Building and Environment*, *46*(10), 1872–1879.

9. Pinzari, F. (2004). Electronic Nose for the Early Detection of Moulds in Libraries and Archives. *Indoor and Built Environment*, *13*(5), 387–395.

10. Portnoy, J. M., Barnes, C. S., & Kennedy, K. (2004). Sampling for indoor fungi. *Journal of Allergy and Clinical Immunology*. Rao, C. Y., Burge, H. a., & Chang, J. C. S. (1996). Review of Quantitative Standards and Guidelines for Fungi in Indoor Air. *Journal of the Air & Waste Management Association*, *46*(9), 899–908.

11. Smith, R. D. (n.d.). Fumigation Dilemma: More Overkill or Common Sense? Retrieved April 06, 2014, from http://www.weito.com/fumigation_dilemma.htm

12. Storey, E., Dangman H, K., Schenck, P., DeBernardo L, R., Yang S, C., Bracker, A., & Hodgson J, M. (2004). *Guidance for Clinicians on the Recognition and Management of Health Effects related to Mold Exposure and moisture Indoors* (p. 70). Farmington: U.S Environmental Protection Agency (EPA).

13. Viitanen, H. (2007). Improved Model to Predict Mold Growth in. *VTT Technical Research.*

14. Wang, Z., Chen, L., & Zhang, G. (2010). Investigation on Indoor Air Quality in University Libraries in Xi'an. *2010 4th International Conference on Bioinformatics and Biomedical Engineering,* 1–4.

15. Weaver-meyers, P. L., Stolt, A., & Kowaleski, B. (1998). Controlling Mold on Library Materials with Chlorine Dioxide: An Eight- Year Case Study, (November), 455–458.

16. Zyska, B. (1997). Fungi isolated from library materials: A review of the literature. *International Biodeterioration & Biodegradation,* *40*(1), 43–51.

Chapter 10

Service Level Agreements for the Digital Library

Masitah Ahmad*, Jemal H. Abawajy

Parallel and Distributed Computing Lab, Scholl of Information Technology, Deakin University, 3216, Victoria Australia

ABSTRACT

Digital libraries offer a massive set of digital services to geographically distributed library patrons. The digital services are commonly sourced from third-party service providers for charge. As externally sourced digital services are becoming prevalence, issues regarding their quality assessment are gaining critical importance. Unfortunately, sourcing digital services from external providers has brought with it stringent quality of service (QoS) demand from the library service users. Currently, there is no way for ensuring QoS between the digital service providers and the library management. In this paper, we propose service level agreements (SLAs) to capture the QoS requirements of the digital service users and the commitments, as well as adherence of the digital service providers.

Keywords: SLAs; Quality of Service; Digital Library; Third-party Sourced Services

INTRODUCTION

Digital libraries have become more prevalent in the library and information science fields. They provide access to digital services in a coherent and economical manner to geographically distributed library patrons. The digital services include desktops, online database, electronic publishing (e-publishing), electronic journal (e-journal), electronic books (e-books), web-portal etc. An advantage of the digital libraries over the conventional libraries is that the former has the potential to store much more information with extremely little or no physical space. Increased accessibility as well as availability to none traditional constituencies of a library for reasons such as geographic location or organizational affiliation is another important advantage of the digital libraries over the traditional libraries. Moreover, digital library users can access some of the digital services from anywhere at any time thus saving their time (Kaur & Singh, 2012).

Digital library has changed the business model from buy-and-use to rent-and-use business model. The advantage of this change is that the libraries will be able to tailor their services to the needs of their current and future users. This in turn, will enable the libraries to be strongly linked to their communities and rapidly adjust to the changing world around them.

As libraries are dedicating increasingly large components of their budget to electronic resources (Plum, et. al, 2010), issues regarding digital services quality evaluation have recently become an area of considerable interest. Several tools such as SERVQUAL to assess service quality within library domains have been developed and widely used. It has been shown that a considerable mismatch exists between the SERVQUAL dimensions and digital service features (Parasuraman et al., 2005). To address such disparity, Vinagre et. al. (2011) developed a tool called dlQUAL that allow the assessment of service quality. Survey-based instruments mainly focus on digital service users as such the data they collect are user impressions or opinions and it's prone to errors (Plum, et. al, 2010). Several approaches such as COUNTER (Counting Online Usage of Networked Electronic Resources) [www.projectcounter.org] and SUSHI (Standardized Usage Statistics Harvesting Initiative) [www.niso.org/workrooms/sushi] that provide statistics of digital service usage also exist. However, the usage data collected are not systematically linked to the desired level of service performance, nor are the results comparable to other institutions (Plum, et. al, 2010).

Despite the great plethora of studies on service quality assessment for library and information science, only a limited number of academic literatures addressed digital service quality evaluations (Vinagre et. al., 2011). The common threat among the existing tools and approaches is that they are all designed to evaluate the performance of the services after they have been deployed. However, digital services provided by digital libraries often include services that exist outside the physical and administrative bounds of the library. These digital services are often contracted from third-party digital service providers for charge. Therefore, we believe that quality of service assessment for digital services requires including the element of third-party service provider. To the best of our knowledge, we are the first to address the integration of Service Level Agreements (SLAs) in the evaluation of library systems.

SLAs have become a valuable tool to help manage service expectations and monitor quality of service (QoS) attributes of services. In digital library, the specification and management of QoS is necessary to enhance user experiences. QoS represents the parameters that can be used to characterise and assess the functional and non- functional aspects of digital services. Some of these parameters are objective in nature and can be automatically measured, whereas others are subjective in nature and can only be measured through user evaluations (e.g., focus groups). Harris & Rockliff (2003) discussing the scope and contents as well as the role of service agreements in Australian health libraries. Comuzzi et. al (2009) focused on establishing and monitoring SLAs for complex service based systems. The authors use business, software and infrastructures services as a SLA hierarchies spanning through multiple domain and layers of a service economy. The authors applying the framework to industrial use cases. However, the proposed SLAs framework specifically on the service provider side only. Therefore, the approach is not suitable for digital library QoS measuring where QoS in the library is also expressed by parameters that focus on the interactive relationship between the libraries with the people whom it is supposed to serve (Hernon & Altman, 2010).

Alhamad et. Al (2010) proposed an approach for SLA framework in cloud computing. The authors use non- functional requirements of services such as availability, scalability and response time to define the SLA parameters for each type of cloud service (Infrastructure as a Service, Platform as a Service, Software as a Service and Storage as a Service). However, the above work is not in direct with the context of SLAs in Digital Library. Moreover, the services in this framework are focusing only toward cloud computing environments. Apart from the nominal work of Harris & Rockliff (2003), creating and implementing service level agreements in libraries does not exist in the published literature. Thus, an approach that guarantees the expected quality of digital services prior to their deployment as well as after they deploy is necessary. In the digital library settings, service level agreements (SLAs) are enormously beneficial if libraries are to achieve their stated mission of serving their patrons. However, there are no academic articles that address SLAs as a tool to create a level of digital service quality in Library. Therefore, the study of such a method is suitable and relevant to be considering because it is increasing in the Digitalenvironments

In this paper, we propose a three-pronged approach for the assessment of digital services quality within a library domain. At the service provider and library interface level, service level agreements (SLAs) are used to establish the required level of digital service quality. At the user-library interface level, the library management collects user experiences and perception through various existing instrumentation. Digital service usage data is collected and used in conjunction with the data that had been collected at the user-library interface. This is to enhance user experiences as well as gauge changes to the level of QoS required. The first two level assessments are also used to validate SLAs. The proposed approach is generic and can be applied to all types of libraries that have standalone digital services or provide integrated traditional and digital library services. We believe that the proposed approach provides valuable performance information to digital libraries' decision-makers, and it can assess digital service quality offered by digital libraries to their stakeholders.

The rest of the paper is organized as follows. A high level conceptual framework is presented in Section 2. Various components of the framework are also described. In Section 3, the proposed SLAs framework for third- party sourced services is explained. Finally, the conclusion and future directions are presented in Section 4.

CONCEPTUAL FRAMEWORK

The digital library provides $\mathbb{S} = \{S_1, S_2, \ldots, S_n\}$ different services (e.g., desktops, e-journals, storage and electronic books) to $\mathbb{U} = \{U_1, U_2, \ldots, U_z\}$ libraries service user (LSU). The digital services can be dedicated (e.g.,hardware such as desktops) or shared (e.g. databases). Also, each service will have a set of attributes such asservice availability that can be quantifiable measures. The library service provider (LSP) contracts the $\mathbb{S}$ servicesfrom $\mathbb{P} = \{P_1, P_2, \ldots, P_m\}$ digital service providers (DSP) for charge. LSP also provides value-added services suchas searching and presentation of information of interest to the LSU. In addition, the LSU is responsible for theplanning and provisioning of the digital services within the

Library. They ensure that content selections,purchasing licensing and access arrangements are in place and understood by the user community. In addition, theLSP will develop a set of KPI to fulfill its clients' need.

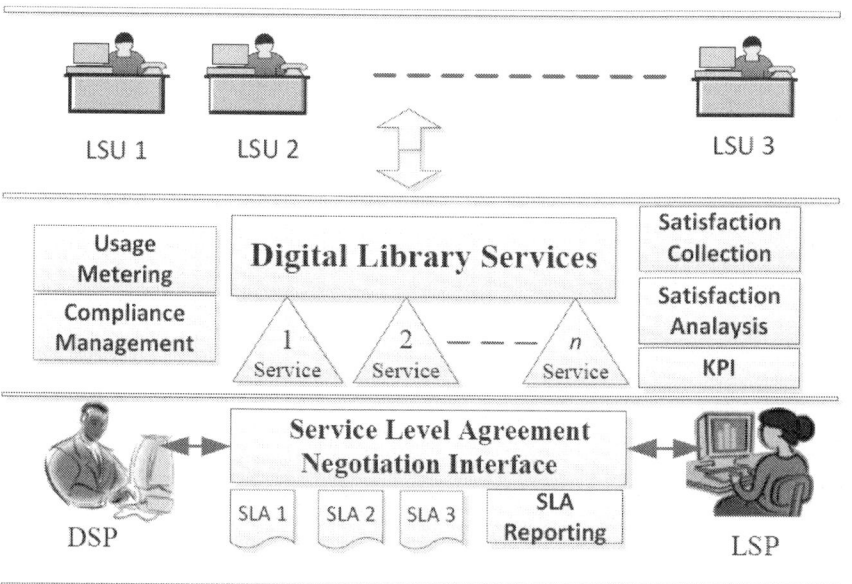

Figure 1: Conceptual framework

Service Level Agreements (SLAs) are core to the relationship between the digital library service delivery functions and the end-users of the digital services. They capture the mutual understanding and commitment of the DSP and LSP regarding the service quality requirements and expectations. DSP and LSP will use SLAs Negotiation Interface to negotiate and establish mutually acceptable agreement on the delivery of the service. SLAs 1, SLAs 2 and SLAs 3 contain the terms of the service level agreements as understood by both the DSP and LSP. It covers items such as the responsibilities of each party (including acceptable performance parameters with applicable metrics), a statement on the expected duration of the agreement, a description of the applications and services covered by the agreement, procedures for monitoring the service levels, a schedule for remediation of outages and associated penalties, and problem-resolution procedures. Measuring and reporting SLAs compliance are the core components of any SLA-based system. SLA reporting is vital for both LSP and DSP as it is one of the stages in SLAs process that indicates the level of compliance. KPIs and SLA metrics are used to measure and assess the digital services performance under SLAs reporting. The report serves as the basis of intervention, validation, justification and direction for agreed SLAs. Hence, SLM reporting is vital for both customer and service provider.

2. SLA-BASED DIGITAL LIBRARY QUALITY EVALUATION

Service Service Level Agreements (SLAs) have become a valuable tool to help manage service expectations and monitor quality of service (QoS) attributes of services in various domains. QoS may contain many metrics that define the deliverables acceptance criteria or serve as standalone measurements of a single aspect of the delivered service. The aims of SLAs are to implement a framework that adapts to changing business priorities and service levels, define clear objectives to shape the service offered by the provider. Effective SLAs not only ensure the delivery of negotiated service quality, but also serves as an efficient service planning and prediction or adjustment processes. Therefore, properly establishing SLAs is crucial to its successful outcome or otherwise.

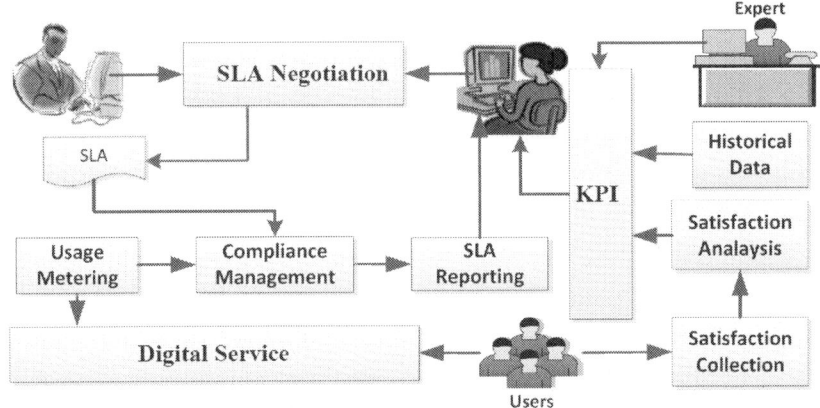

Figure 2: SLA-based digital service quality evaluation framework

In digital library, the specification and management of QoS is necessary to enhance user experiences. QoS represents the parameters that can be used to characterise and assess the functional and non-functional aspects of digital services. Some of these parameters are objective in nature and can be automatically measured, whereas others are subjective in nature and can only be measured through user evaluations (e.g., focus groups). The proposed SLA-based approach is designed to move away from subjective measures based on opinions. Harris & Rockliff (2003) discussing the scope and contents as well as the role of service agreements in Australian health libraries. Comuzzi et. al (2009) focused on establishing and monitoring SLAs for complex service based systems. The authors use business, software and infrastructures services as a SLAs hierarchies spanning through multiple domain and layers of a service economy. The authors applying the framework to industrial use cases. However, the proposed SLAs framework specifically on the service provider side only. Therefore, the approach is not suitable for digital library QoS measuring where QoS in the library is also expressed by parameters that focus on the interactive relationship between the libraries with the people whom it is supposed to serve (Hernon & Altman, 2010).

Fig. 2 shows the digital services quality evaluation framework. The purpose of the framework is to achieve expected level of digital service quality, through a periodic cycle of negotiation, agreeing, monitoring and reporting upon delivered digital service. The library benefits from a clearer picture of the library users expectations, the ability to balance and adjust their resources to meet those expectations, as well as explicitly detail the costs associated with any given level of service. To achieve these aims, the library management develops a set of KPIs that dictate what is important to the library clients and the librarians using historical performance data and expert knowledge. The main idea here is to convert both subjective and objective data collected by the librarians using the conventional methods into KPI's. These KPIs define the expectation of the level of service, which the library patrons can expect to receive and specified in terms of an achievable service level. KPIs are specific measurable characteristics of the digital services such as throughput, availability, response time, or quality of support. KPI for the service must accurately reflect the expectations and perceptions of both the service user and service provider. It should also be directly linked to a value, which can be consistently monitored.

The next step is to develop SLA in collaboration with the service provider. A consensus between the digital service provider and library service provider on the services delivery is critical for a service agreement to be successful (Harris & Rockliff, 2003). Basically, SLAs are intended to ensure that the service provider understands the expected service quality level they are supposed to deliver, the customer knows what to expect, and both can see what is actually being delivered. Therefore, the library service providers will need to communicate the quality requirements, how it is monitored and measured with the service providers quite succinctly. To this end, the library management will select and start negotiation with service providers based on its KPI's with the aim to reach a service level agreement (SLAs) that ensures high quality and timely delivery of digital services to support the library business. SLAs must capture the mutual understanding and commitment of the digital service provider and the library management regarding the service quality requirements and expectations. Commitments are responsibilities that digital service providers must meet to fulfil service level agreements for agreed amount of remunerations from the library. In order to avoid misunderstandings, SLAs guarantee terms need to be explicitly related to reasonable, attainable performance levels and measurable metrics. Also, SLAs should formally state the exact settings under which the digital services should be delivered. SLAs should ensure that the level of digital service delivery is objectively measured based on KPIs and should also be in compliance with relevant best practice and standards. The SLA should include a provision in which the service provider agrees to assure the library for any breaches of its agreement. Furthermore, SLAs should be clear and simple to ensure that it is possible to determine compliance.

With the SLAs in place, the library patrons are given access to the services. The library management will collect and analyse information related to its client satisfaction level with the service provided through the conventional mechanisms such as surveys. Even though some of the information collected is subjective, they can serve as a check on the validity of the SLAs. On the other hand, the digital service provider will monitor the quality of the service delivered and generate reports. The service provider will also make SLA report available to help the library management to authenticate and oversee the quality of services delivered through scheduled and on-exception reports. The library management can use the internally collected library client's level of satisfaction with the perceived level of service provided and the reports from the service provider to check how that the commitment as specified in the SLAs is faring, whether service levels have been maintained and whether you are owed any rebates for service outages or to renegotiate the terms of the SLAs if need be. We believe that making the SLAs two-sided and by measuring the end users satisfaction on mutually dependent metrics is a good way to concentrate on the intended outcomes.

CASE STUDY

We now illustrate the proposed approach using a networked desktop system provided to the library from a third-party service provider. The motivation for using this example is that people increasingly depend on the local library's public access computers, Internet access, and reference support to search for jobs, take classes, complete homework assignments, obtain medical information, and receive government information and services (Fleischmann, 2010). The various QoS properties such as availability, accessibility, performance, reliability, and security should be addressed in the creation of SLAs.

The library management identify IT services and service requirements and define, build and negotiate Service Level Agreements (SLAs)

1. KPI1: Percentage increase in customer perception and satisfaction of SLAs achievements, via service reviews and customer satisfaction survey responses

2. KPI2: The system (i.e., hardware, software, and network) must be functioning and available 99% of the time during business hours.

3. KPI3: Customer support for service maintenance requests must not exceed 12hours at most.

KPI1 is for the sole purpose of use by the library. KPI2 dictates system availability guarantees over a period of time. KPI3 includes the typical help desk problem reporting and problem resolution guarantees based on severity   level. Severity level and response and resolution times are assigned according to their impact on customers. The   acceptable response time and resolution time are negotiated between the IT Service Provider and the Customer.

For the second and third KPIs, the library management will negotiate and develop an SLAs with the service   provider. The developed SLAs will specify that if the system fails to meet the negotiated 99% uptime, the library is   entitled to reduce its bill by an agreed-on percentage. For instance, if the system is unavailable for an hour, the   library is entitled to a 10% rebate of its monthly service fees; in the case of a service outage for two hours, the library is entitled to a 20% rebate of its monthly service fees" and so on. The SLAs also describes the procedures   for reporting any problems with the service to the service provider; notifying library management about all   scheduled maintenance as well as generating SLAs reports and on-exception reports. It will also include scope for   renegotiation and meeting response and resolution times associated with service related incidents. For example, for   KPI3, average speed of answer (e.g., 15 seconds), target service level of 95^ calls answered in 15 seconds, and average talk time of less than 3.5 minutes per call can be stated.

The library management will collect satisfaction data through its own instrumentation to measure the level of   KPI1 achievement. This is essential for the library to see if the digital service provided capacity is below or above   that needed to meet the clients' needs and adjust the service accordingly. This will require renegotiation of the   SLAs.

4. CONCLUSION AND FUTURE DIRECTIONS

The assessment of digital services is a key element in the delivery of digital library services to meet the needs   of the library users. In this paper, we have argued that SLAs are principally valuable for correlating library patron   experience metrics with the underlying infrastructure components that support the associated business service. We proposed a three-pronged approach for the assessment of digital services quality within a library domain. At the service provider and library interface level, service level agreements (SLAs) are used to establish the required level of digital service quality. At the user-library interface level, the library management collects user experiences and perception through various existing instrument. At the user-service interface level, the digital service usage data is collected and used in conjunction with the data collected. Due to that, the user-library interface is to enhance user experiences as well as gauge changes to the level of QoS required. The first two level assessments are also used to validate SLA. The proposed approach is generic and can be applied to all types of libraries that have standalone digital services or provide integrated traditional and digital library services.

ACKNOWLEDGEMENTS

The research leading to these results has received funding from the parallel and distributed computing laboratory at Deakin University and the Ministry of Higher Education of Malaysia (MOHE) through its sponsorship on PhD program.

REFERENCES

1. Alhamad, M., Dillon, T. & Chang, E. (2010). SLA-Based Trust Model for Cloud Computing. Network-Based Information Systems (NBiS), 13th International Conference on , vol., no., 321-324.

2. Cecilia Garibay, Humberto Gutierrez & Arturo Figueroa. (2010). Evaluation of a Digital Library by Means of Quality Function Deployment (QFD) and the Kano Model. The Journal of Academic Librarianship 36 (2), 125- 132.

3. computing." In J.C. Bertot, P.T. Jaeger, & C. McClure (Eds.), Public Libraries and the Internet: Roles, Perspectives, and Implications, (91-102). Santa Barbara, CA: Libraries Unlimited.

4. Comuzzi, M., Kotsokalis, C., Spanoudakis, G. & Yahyapour, R. (2009). Establishing and monitoring SLAs in complex service based systems, in: Proceedings of the. IEEE International Conference on Web Services, IEEE Computer Society, 783–790.

5. Fleischmann, K. R. (2010). The public library in the life of the Internet: How the core values of librarianship can shape human-centered

6. Harris, L. & Rockliff, Sue. (2003). Implementing Library Service Agreements: The Experience of Australian Health Libraries.10th Asia Pacific Special Health and Law Librarians Conference. Adelaide. Retrieved from http://conferences.alia.org.au/shllc2003/papers/004.pdf.

7. Hernon, P. & Altman, E.(2010). Assessing Service Quality: Satisfying the expectations of library customers. American Library Association, Chicago, IL.

8. Jun Woo Kim & Sang Chan Park. (2010). Outsourcing strategy in two-stage call centers. Computers & Operations Research. 37 , 790-805. Kaur, K. & Diljit, S.(2012). Modelling Web-based library service quality. Library information Science Research, 34 (3),184-196.

9. Nor Irvoni Mohd. Ishar and Mohd. Saidfudin Masodi. (2012). Students' Perception towards Quality Library Service Using Rasch Measurement Model. Innovation Management and Technology Research (ICIMTR), International Conference on, vol., no.,668-672.

10. Parasuraman, A.,Valarie A. Z. & Arvind M. (2005). E-S-QUAL: A Multiple-Item Scale for Assessing Electronic Service Quality. Journal of Service Research, 7(3), 213–3.

11. Plum, T., Franklin, B., Kyrillidou, M., Roebuck, Gary. & Davis, M. (2010) "Measuring the impact of networked electronic resources: Developing an assessment infrastructure for libraries, state, and other types of consortia", Performance Measurement and Metrics, 11(2),184 -198.

12. Shang Gao, John Krogstie & Keng Siau. (2011). Developing an instrument to measure the adoption of mobile services. Mobile information systems, 7 (1), 45-.

13. Vinagre, M. H., Leonor G. P. & Paula O.(2011). Revisiting digital libraries quality: a multiple-item scale approach. Performance Measurement and Metrics, 12(3),214–236.

An Exploratory Study of Collaboration in New Zealand Tertiary Libraries

Colleen Finnerty[a]

ABSTRACT

The shift in policy from market driven behaviour towards a morecooperative tertiary sector is having an effect on New Zealand academic libraries and their relationships. Despite this, there has been no investigation of collaboration specifically targeting New Zealand tertiary libraries. This research project examine the state of collaboration between New Zealand tertiary libraries in early 2004. Its objective was to explore the extent and nature of collaboration, and the attitudes of New Zealand librarians towards this process. The research found that the majority of tertiary libraries are collaborating (88%) in some form with three types of collaboration dominating the results: joint licensing agreements (20%), reciprocal borrowing (20%), and acquisition purchasing agreements (1 9%). These ventures are initiated by directives from the libraries' own institutions, or by the formal and informal gathering of librarians where collaboration was used to solve a variety of problems. Once initiated, these ventures are often informally constructed, with only I 0% having a written policy and 22% having a written contract. Despite identified barriers such as a lack of resources, and the need to give priority to local user needs, respondents (79%) felt that collaboration would continue to increase from its present rate.

In New Zealand during the 1980s and into the 1990s, the public sector, including tertiary institutions and their associated libraries, was faced with decreased funding and government policies based on competitive business principles. Consequently, there were no major new collaborative initiatives other than the computerisation of the Te Puna interloan scheme. A change to a Labour government in 1999 signalled a shift from market driven policy towards a more collaborative approach to support New Zealand in becoming a knowledge society. A knowledge society should allow information to flow smoothly to fulfil New Zealand's national goals of economic and social transformation. The Labour government's Tertiary Education Strategy, 2002-2007, focused on providing a blueprint for a collaborative tertiary system that contributed to these national goals by providing research, information access, and relevant skills for a changing society. To achieve these goals the tertiary sector has been encouraged to shift from isolation to connectedness with others. The combination of government policy and information technology that allows for the rapid transfer of information between institutions and individuals has meant that tertiary libraries have also embraced the move to closer relationships with those around them. Although few libraries would argue that they are any more affluent than they once were 'there is evidence of a growing feeling in New Zealand libraries that partnerships or some form of collaboration are required to achieve the level of services that users are expecting.' 1 This has encouraged and reinforced the collaborative nature of New Zealand tertiary libraries which have always had an ethical commitment to sharing, in recognition that no library can be all encompassing.

LITERATURE REVIEW

Reasons to Collaborate

While there are many published articles dealing with library collaboration, mostare of a descriptive nature explaining how resource sharing was established and why in a particular northern hemisphere locality. For example Allen and Hirshon2 surveyed members of the International Coalition of Library Consortia (ICOLC) to establish why libraries are being driven to form collaborative ventures and what possibilities cooperation presents for the future. One of the recurring reasons was the need to cut costs and be seen to be competitive to outside authorities, in times of reduced funding and increasing competition from other information providers. Changes in information access and delivery, and the rapid growth of information technology have also been cited as reasons for the formation of collaborative ventures. Libraries can now form networks to hare electronic as well as physical items and negotiate joint licensing agreements that are more favourable to the libraries. Potter3 believes that with the information explosion libraries cannot hope to house all needed information, and so collaboration acts as 'recognition of the fact that a group of libraries has a combined set of resources that is greater than the resources of a single member.' In exploring collaboration, many hope to increase the amount of information to which their patrons have access, with moves away from physical ownership to access to information in various distributed forms, such as electronic journals.

Barriers to Collaboration

Evans4 explains there are 'seldom reports on how and why a cooperative project failed' and indicates that this lack of knowledge about past efforts is a barrier to collaborative success. What the literature does indicate as one of the main reasons for failure is the inability of libraries to cede some control. Evans5 points to fear of decisions being made that individual libraries must agree to, and fear of budgets being directed away from core users. Few if any academic libraries would argue that collaborative demands should be placed above primary users' priorities and their core collection. It is difficult to develop arguments of cooperative collection development success or failure when there is little evidence to support either assertion due to a lack of objective measurements which is part of a larger difficulty that collection development has measuring its effectiveness in either qualitative or quantitative terms.

This lack of evidence is apparent again when looking for documented savings to acquisition budgets. Shoaf believes the effectiveness of collaboration is based on 'qualitative measures that when related to budgets were based on overall library expenditure and could not be broken down by areas.'

Critical Success Factors

With the emphasis on maintenance of local collections and control, it comes as no surprise to see that one of the most heavily cited critical success factors in collaboration is that 'programmes must be responsive and minimally threatening to local priorities and programmes. '7 To ensure local needs are met, while still achieving collaborative goals, Allen and Hirshon8 argue that members must have a 'high degree of respect for and recognition of the value of increased collaboration,' and also strong leadership. Giordano9 sees the availability of funding to establish a centralised authority with dedicated staff as an ideal way to efficiently accomplish collaborative goals. Despite the need for autonomy, Allen and Hirshon10 denote a 'direct correlation between how much autonomy a member retains and the success of the consortium- the less formal a consortium the less likely its goals will be successfully achieved.' Clayton and Gorman 11 also support a formal governing structure, which has the responsibility to issue directives and review procedures, such as the consortium agreement, which can also outline ways to share costs and develop policies.

The Nature of Collaboration

The variation in the forms that collaboration takes is illustrated by the variety of models that are presented in the literature. Sinclair12 was the first to propose four models of collaborative activity among libraries that Gorman and Cullen 13 still consider 'as a valuable guide today.' The first is the bilateral exchange model, in which two participating libraries exchange materials such as reciprocal borrowing agreements. The second is the pooling model, where more than two libraries contribute to a draw from a common pool. The dual-service model is the third which involves two or more libraries taking advantage of the facilities of one of the participants to produce a common output such as a shared OP AC(Online Public Access Catalogue). The last type of collaborative activity is referred to by Sinclair14 as the service centre model, where a number of libraries employ the services of a facilitating organisation to input and process materials for the individual libraries, rather than for common output. OCLC is an example of this model. Recently Allen and Hirshon 15 have developed a guide that places each venture on the point of a continuum based predominantly on the governance structure, from a loosely knit federation at one end, to a centrally funded state-wide consortium on the other. This model takes into consideration the variety of informal arrangements that exist. Other models have been based around how collaborative programs have been funded (O'Connor)16 or, by sector and specific interest, such as a particular discipline as described by Clayton and Gorman. 17

Theories behind Collaboration

Hayes18 believes games theory and its cooperative games are useful to apply to collaborative ventures, as this theory recognises 'bargaining is a process of making offers and demands, with the objective of achieving total joint results that are better than what can be obtained from simply the competitive.' Underlying any economic theories, though, is what Hayes19 describes as the 'ethical commitment of librarianship to the very concept itself, in recognition that no library can be all encompassing and that sharing is the only way to ensure preserving the record of the past, and providing access to that record.'

New Zealand Libraries in Collaboration

Literature on collaboration in New Zealand tertiary libraries is very sparse. Information is largely descriptive in nature exploring the possibilities collaboration could hold, and issues that need to be contended with before implementation. In a descriptive report funded by the National Library of New Zealand Domer and Annear20 have examined the reasons why collaboration should be gaining momentum in New Zealand, what benefits could be derived from these ventures, and what issues stand in the way. Again this research is based largely on information gained from North American experiences. Domer and Annear's report is supplemented by Helen Renwick's21 report to CONZUL (Council of New Zealand University Librarians) and the New Zealand ViceChancellors' Committee, which considered the possible benefits for New Zealand's eight universities undertaking more formal consortia arrangements, and made numerous recommendations on what should be considered. The only statistical survey of resource sharing in New Zealand libraries was a survey in the public library setting ten years ago. Colyer22 has described New Zealand public library collaboration as 'schemes that do not have a written contract or procedures but are characterised by the informal use of existing channels of communication.' To what extent tertiary libraries mirror the public library experience is unknown. Recent developments such as PER: NA (Purchasing Electronic Resources: A National Approach) and LCoNZ (Library Consortium of New Zealand) underline the significance and growth of this phenomenon in New Zealand. However, this movement is largely unexamined by research of any sort from a New Zealand perspective.

The purpose of this survey23 was to see how widespread collaboration is,and with whom. The survey also examined what areas within the library are most affected by these ventures. Is collaboration focused on creating financial breaks for libraries, or for the benefit of the client base? The survey also explored the nature of collaborative relationships by seeking to establish howventures are being initiated and maintained. Who approaches whom when libraries interact in this way? What agreements are made to ensure that collaborative programs progress? The survey also asked respondents where they envisage collaboration heading in the future. Is there a fundamental shift in how libraries interact? If so, what barriers need to be identified and addressed before collaboration can meet its full potential? The survey questions were framed to answer the seven research objectives which are outlined below, followed by the related findings.

METHODOLOGY

A concurrent mixed methods survey collecting both qualitative and quantitative data using both predetermined and multiple option answers was used for this

research. Mixed methods allowed for a comprehensive picture of collaboration,with both statistical information and specific themes drawn from text analysis being integrated into the final results and interpretation. This method also allowed for the triangulation of data collection and analysis to assure the validity and reliability of the research results.

Tertiary libraries were selected for examination for a number of reasons. Firstly, as pointed out earlier in the literature review for this topic, there were nostatistical data exploring collaboration in the New Zealand tertiary library sector. Tertiary libraries were also selected because these libraries are placed firmly within the educational field, where much of government policy is focused to produce and direct the knowledge society. Hence, tertiary libraries are at the forefront of exploring collaborative ventures to meet the objectives of allowing seamless information flow beyond the boundaries of each other's limited resources.

New Zealand Contacts in Libraries and Information Services is a library industry directory which is published in association with the Library and Information Association of New Zealand Aotearoa by Contacts Unlimited. This directory was used to identify all 68 tertiary libraries in New Zealand, and their respective managers. The survey was aimed at library managers, as they were considered key informants because of their responsibilities concerning the operational and strategic aspects of library management. By the cut-off date 19 April 2004, 33 surveys from a total of 68 had been returned. The response rate of 48% was slightly less than the targeted response rate of 50%. However, two respondents from libraries with branches chose to return only one survey on behalf of the whole institution. This reduced the population size from 68 to 65, which increased the response rate to pass over the desired 50% return rate. The majority of participants identified themselves as working in management positions (8 8%), and in either a university (41%) or polytechnic (41%) library. Once the surveys were returned, each possible quantitative response was assigned a code. This allowed the information to be added to an Excel spreadsheet to aid in establishing descriptive statistics and presenting them in graphs. For the qualitative data gathered in response to open ended questions, a coding manual was devised to identify the main themes using content analysis.

FINDINGS

Objective 1: To discover the extent of collaboration in New Zealand tertiary libraries

The extent of collaboration was considerable with nearly 88% of libraries involved in collaborative ventures over and above the Te Puna interloan scheme.Respondents stated that most ventures were undertaken with other tertiary libraries. However, over 31% of respondents mentioned collaborative projects with other types of information providers such as public libraries and Crown Research Institutes. This is important as it shows that tertiary libraries are looking outside the traditional tertiary sphere for partners.

Objective 2: To determine what areas of library operations are most affected by collaborative venture'

Three types of collaboration dominated the results: joint licensing agreements (20%), reciprocal borrowing (20%), and acquisition purchasing agreements (19%). These results as displayed above in Table 1 show libraries seeking to increase the amount of print and electronic resources available to their users while controlling costs. Collaboration is one way libraries are achieving this goal. Five respondents acknowledged a shared library system. Although it is not possible to conclude that all respondents are from LCoNZ (Library Consortium of New Zealand) libraries, it can be concluded that the focus of library cooperation has extended beyond purely book buying groups. This mirrors research undertaken in the northern hemisphere which shows that collaborative projects that were once formed to physically share materials are now creating networks and sharing electronic resources.

Table 1

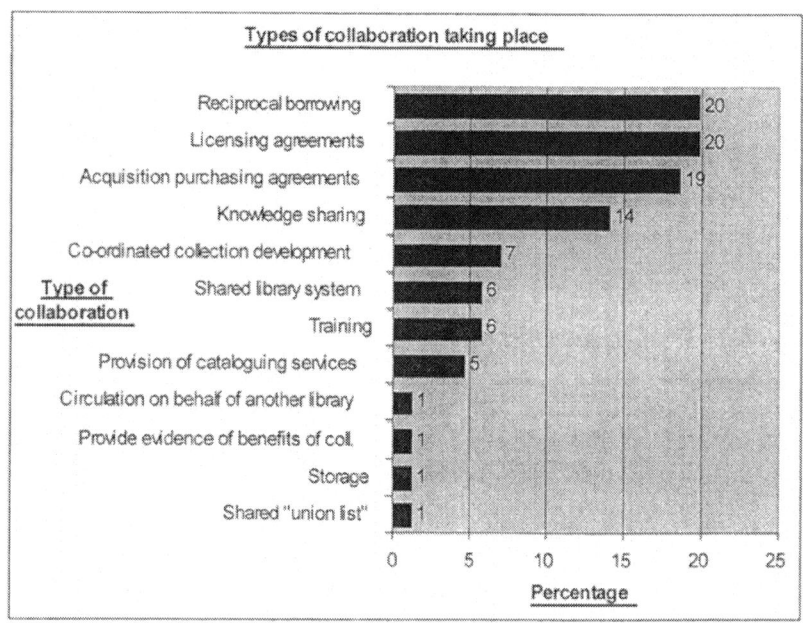

Types of collaboration taking place

Objective 3: To explore how collaborative ventures are initiated and maintained

One third of respondents mentioned directives from their parent organisation as instrumental in establishing collaboration. Formal gatherings of information · professionals were another vehicle for collaboration. These forums allowed library professionals to identify common problems which have led to joint solutions. More informal networks were also found to be important. Over a third of respondents indicated that personal contact with another librarian was instrumental in instigating collaboration. Libraries would identify a requirement, such as the need for cataloguing training, and approach other institutions they knew to be capable of meeting that need. This informal network of familiarity amongst librarians can be seen as having a large impact on how initiatives are started. All three of these scenarios would have been less likely to influence collaboration if it had not been for a change in governmental policy. The push for collaboration rather than competition at government level has had a direct impact on library operations.

Once initiated, these ventures are often informally structured and maintained purely by staff communication between institutions. When participants were asked to discuss their latest collaborative venture only one quarter of ventures had staff delegated to manage and administer them. In addition to that less than 14% had allocated funding. The ventures were largely without written contracts, with only 22% having a contract and less than 10%

having a written policy. However, over 24% of respondents indicate regular

meetings. This is in stark contrast to the northern hemisphere where a lack of funding and formality is associated with a lack of success.

Objective 4: To identify the reasons libraries are undertaking collaborative ventures and which of these is most important

Table 2

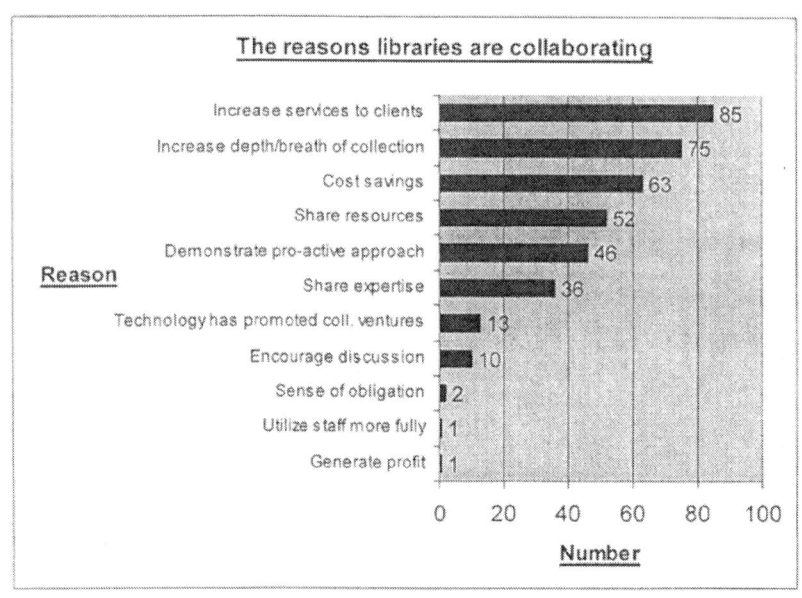

The reasons libraries are collaborating

Reason	Number
Increase services to clients	85
Increase depth/breath of collection	75
Cost savings	63
Share resources	52
Demonstrate pro-active approach	46
Share expertise	36
Technology has promoted coll. ventures	13
Encourage discussion	10
Sense of obligation	2
Utilize staff more fully	1
Generate profit	1

The most important reasons for libraries to collaborate are displayed in Table 2. The most important overall reason was the need to increase 'services to clients' followed in second place by the 'need to increase the depth and breadth of the collection.' Ranked third was the 'need to save costs.' These top three motivations behind collaboration in New Zealand indicate that libraries are trying to improve resources and services while reducing costs. Although this is hardly a revelation, it points to the fact collaboration is being used as a means to achieve this end.

Another reason which ranks highly with the need to save costs and share resources is the wish to demonstrate a proactive approach to library services. All libraries surveyed are part of a larger organisation to which they are accountable. Despite this need to be accountable, the results do not indicate 'generating profit' as a motivation for collaboration. Librarians hold many marketable and saleable skills, but their emphasis remains altruistic.

Objective 5: To find out what librarians consider to be the critical success factors in establishing library collaborative projects

Table 3

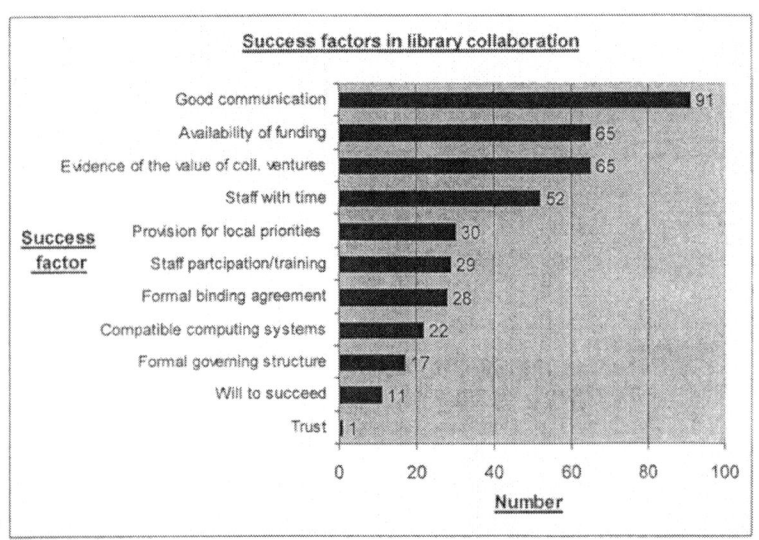

'Good communication' was strongly identified as the most important critical success factor in collaboration. Tied for second equal were the factors 'availability of funding' and 'evidence of the value of collaborative ventures.' The third most important factor was 'staff with time.' These four elements dominated the list of critical success factors as displayed in Table 3. This is surprising given other research findings from the northern hemisphere where the acknowledgement and provision for local priorities, and the need for a formal binding agreement or governing structure, were more visible in research results.

One of New Zealand's most recent collaborative ventures is EPIC (Electronic Products in Collaboration), which involved numerous types of libraries, including tertiary libraries. They established a governance group as part of the venture, to make recommendations on funding, governance and a structured partnership model. This was in reaction to the understanding that if EPIC was to succeed, it needed to acknowledge and act upon a variety of concerns. Very few collaborative projects in New Zealand are as wide reaching and multi-faceted as EPIC, but it is an example that mirrors experiences in the

northern hemisphere. The results from the survey indicate, however, that most New Zealand tertiary libraries' collaborative ventures are less formalised.Perhaps this is why communication was ranked so highly as most ventures were based on goodwill. However, it is true that many of New Zealand's tertiary library ventures are still in their infancy. They are also on a smaller scale than those of the northern hemisphere and thus may not require the same level of formal structure.

Objective 6: To determine what barriers need to be addressed if collaboration is to be successful

Table 4

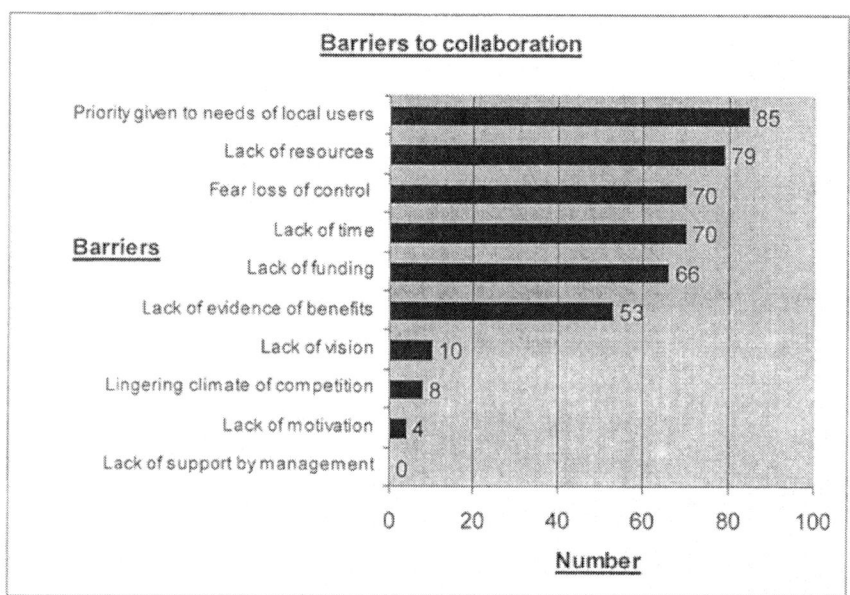

Respondents felt that their need to 'give priority to the needs of local users' was the biggest barrier to collaboration. Second was the 'lack of available resources' and equal third was a 'lack of time' and the 'fear of losing control over library resources and policy.' Overseas research supports the notion that libraries find it difficult to cede control. There is a fear of decisions which individual libraries must agree to, and fear of budgets being directed away from core users. It was apparent in the survey that these fears are also shared by New Zealand libraries, with 'priority being given to the needs of local users' being the most important barrier to collaboration and 'fear of the loss of control over library resources and policies' being third. Could these fears be allayed by a written policy or binding contract that could protect and set out individual institutions' interests while exploring the benefits of collaboration?

Respondents continue to point to a lack of resources, both financial and in staff time, to dedicate to collaboration as presented in Table 4. There appears to be a conflict in interest between librarians being able to serve their own clients, as well as working for the common good. Respondents felt that often there were not enough resources to go around to serve their own users, let alone to dedicate time and resources to other ventures. Perhaps these problems are not helped by the fact that the cost of collaboration is immediate while benefits may be more gradual. If collaboration is to be successful it appears two things must happen. First, libraries must be persuaded of the benefits of collaboration. Second, libraries must be willing to dedicate resources and time to the venture.

Objective 7: To gauge where New Zealand tertiary libraries see collaboration heading in the future

Over 78% of respondents felt that collaboration would increase from the present rate. This suggests that while the path to collaboration is challenging, the majority of libraries see it as a vital part of their future. Over 90% of respondents agreed that there has been a fundamental shift from a self sufficient model to a more outwardly focused model. This has been influenced by government support for a more fluid tertiary sector which requires a move from a competitive to a more collaborative climate. Technology has also acted as a catalyst for increasing collaboration by not only making it possible, but by making it an imperative due to rising costs. Visible successes will act as incentives for further collaboration in the future.

CONCLUSION

In conclusion the natural inclination of tertiary libraries to collaborate must be encouraged. New Zealand is a small country with a tertiary library sector constrained by limited resources and funding in an era of information explosion and technological revolution. Tertiary libraries without the restrictions of divisive competitive principles can overcome these restrictions by pooling resources and negotiating as a group. This survey has pointed out that there is a belief in collaboration as a continuing force in libraries. With an environment ripe to foster the growth of collaboration, libraries can use this belief and process to reach their potential for their users and maintain relevancy in a networked world.

ACKNOWLEDGEMENTS

I would like to express my gratitude to Kat Turner and Brenda Chawner, who supervised this research project and encouraged me to write this article.

NOTES

1. D G Dorner and J Annear The Renaissance of Library Consortia: Implications for New Zealand Libraries: A Report Commissioned by the National Library of New Zealand Wellington National Library of New Zealand 2000 p2

2. B McFadden Allen and A Hirshon 'Hanging Together to Avoid Hanging Separately: Opportunities for Academic Libraries and Consortia' *Information Technology and Libraries* vol 17 no 1 1998 pp36-45

3. W G Potter 'Recent Trends in State Wide Academic Library Consortia' *Library Trends* vol45 no 3 1997 p416

4. G E Evans *Developing Library and Information Centre Collections* Libraries Unlimited Inc 2000 p474 *Ibid*

5. E Shoaf 'The Effects on Consortia Membership on Library Planning and Budgeting' *Library Administration and Management* vol 13 no 4 1999 p200

6. P H Mosher and M Pankake 'A Guide to Coordinated and Cooperative Collection Development' *Library Resources and Technical Services* vol 27 no 4 1983 pp 425 McFadden Allen and Hirshon 'Hanging Together. .. ' p43

7.   T Giordano ' Library Consortium Models in Europe' *Alexandria* vol 14 no 1 2002 pp41-52

8.   McFadden Allen and Hirshon 'Hanging Together ... ' p44

9.   P Clayton and G E Gorman *Managing Information Resources in Libraries: Collection Management in Theory and Practice* London Library Association Publishing 2001

10. M P Sinclair 'A Typology of Library Cooperatives' *Special Libraries* vol 64 no 4 1973 pp181-186

11. G E Gorman and R Cullen 'Models and Opportunities for Library Cooperation in the Asian Region' *Library Management* vol21 no 7 2000 pp373-84 at 378

12. M P Sinclair 'A Typology ... '

13. McFadden Allen and Hirshon 'Hanging Together. .. '

14. S O'Connor *Beyond Cooperation into Consortia! Approaches.* Conference paper delivered to *Information Online and on Disc 99 Conference* 19-21 January 1999 Sydney. Available at http://www.caval.edu.au/about/staffpub/paper4_ soO 199 _publications.html

15. Clayton and Gorman *Managing Information Resources in Libraries*

16. R M Hayes 'Cooperative Games Theoretic Models for Decision-Making in Contexts of Library Cooperation' *Library Trends* vol 51 no 3 2003 pp441-462

17. *Ibidp455*

18. Dorner and Annear The Renaissance of Library Consortia

19. H Renwick The Big Picture: Opportunities for Potential Areas for Closer Collaboration Among the New Zealand University Libraries. A Report to CONZUL and the New Zealand Vice-Chancellors Committee. Available athttp://www.conzul.ac.nz/BigPicture.pdf

20.   S Coyler 'Effective Resource Sharing in New Zealand Libraries' *New Zealand Libraries* vol47 no 10 1994 pp 193

21.   C Finnerty Collaboration in New Zealand Tertiary Libraries - an Exploratory Study unpublished research project for the MLIS at Victoria University of Wellington 2004

Chapter 12

The Adaptable Cycle of Engagement: A Win/Win Model for the Library

Sheryl Kaye

ABSTRACT

As libraries face greater financial challenges while redefining their roles, public engagement has been put forth as one way to address both concerns. In fact, such efforts have produced solid results for many libraries, even when not following a clear path with a well-defined outcome. Administrators believe in the concept, but sometimes lack a foundational understanding of how public engagement works. This paper's author spent several years as a published journalist and business consultant observing and analyzing the ways in which companies utilize public engagement and the manner in which these campaigns resulted in tangible profits. The author then developed a paradigm called the Adaptable Cycle of Engagement (ACE) in order to offer a model for successful implementation of such an agenda. This paper defines that system in detail, showing step by step how public engagement works to ultimately bond the community to the library, enhance library offerings to patrons, and enable the library to reap financial rewards.

Keywords: *Library; Public Engagement; Social Media; Budgets; Return on Investment; Community; Patrons; Bonding; Progressive Steps; Adaptable Cycle*

INTRODUCTION

Public or community engagement is critical to the success of the library. In fact, an important strategic advocacy objective of the American Library Association is to "increase public awareness of the value and impact of all types of libraries and the important role of librarians and other library staff." (American Library Association, 2013, A.1.6, strategic objective 1). ALA has also partnered with the nonprofit Harwood Institute for Public Innovation to create a program called *Libraries Transforming Communities* (LTC), whose goal is to help libraries move from an internal paradigm (being library centric) to an external one that focuses on interaction with community and the potential positive results that are gained when this occurs (American Library Association, n.d.). It is only recently that most libraries have started to embrace such a focus. While most library staff can define public engagement, many employees report no involvement in the development of such programs, have little to do with delivering the programs, and receive little or no training in how to actually engage the public (Museums, Libraries, and Archives, 2011, p. 5).

Having spent time in the for-profit business arena as a published print and photojournalist and business consultant, I have observed and documented numerous companies that have utilized community engagement and social media tools to increase visibility in a competitive marketplace and to generate revenue. Although the public library is not a for-profit institution in America, the techniques that corporations use to successfully implement public engagement can also apply to the library. After searching for an applicable roadmap, however, I found no formal theoretical model structured with the library in mind. Hence the development of the Adaptable Cycle of Engagement (ACE) presented below and in Figure 1. Each "spoke" of the Adaptable Cycle of Engagement helps propel the sequence of public engagement to the next phase. The ultimate goal is a public that is more invested in the library, leading to increased support.

CONCEPTUAL BUILDING BLOCKS OF THE MODEL

The first step in the ACE model is for a public library to tangibly connect with its community (see Figure 1). Fortunately, technology offers libraries many ways to share information with their communities and initiate engagement. Tools like Facebook, YouTube, Twitter, Instagram, Google+, Pinterest, and e-newsletters enable the library to reach out to a large number of people, sometimes in a matter of minutes (Dankowski, 2013, p. 39). For example, by employing its social media, the New York Public Library had success in informing the local community on a variety of topics, including the benefits of membership, so much so that after a one-year social media campaign, the Library had increased new card sign-ups by 21 percent (Eckerle, 2013).

In addition to social media tools, public libraries can host public events, offer in-house training, and exhibit art and other cultural objects, all in an effort to not only tender services to its patrons, but also communicate in new ways. Whatever technique is employed, the library is initiating the conversation and then continuing it as an active participant in the library-patron relationship.

This leads to the second step, where library users become better informed and educated about the institution and all that it has to offer. Studies show enhanced data leads to a deeper and more profound understanding of any concept (Finley, 2000; Dumais, Cutrell, & Chen, 2001). The more one absorbs about anything, the more perceptive one is on that topic. The library is no longer a general concept to patrons, but rather an accessible and comprehensible resource.

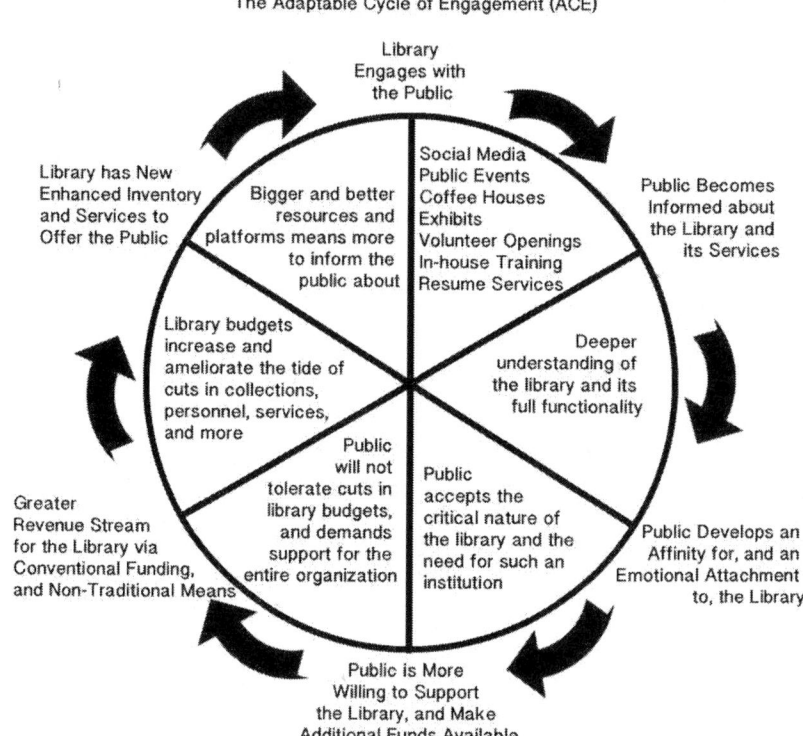

Figure 1. Adaptable Cycle of Engagement

This strengthened perception among patrons is significant because it leads to the next stage in the ACE model: once members of the community understand the role and services provided by the library, they develop a heightened affinity for the

institution. Then an emotional attachment to, and a passion for, the library can begin (Wooden, 2006, p. 7).

For example, the Vancouver Public Library (VPL) system in Canada established such a connection when it employed social media via a Twitter campaign. After only one year, VPL garnered more Twitter followers than any other system in Canada, opening up opportunities for immediately engaging the community, including broadcasting emergency information, offering general customer service, and enabling instantaneous patron feedback (Cahill, 2011, p. 265). Library administrators reported that with this tool, and similar social media platforms, the VPL system built a powerful and enduring relationship with the community (Cahill, 2011, p. 268).

There is no shortage of reports of libraries facing budgetary challenges in the past five years, often resulting in staff reductions, cuts in services, and curtailed collections (Henderson & Lonergan, 2011, p. 7). Of even greater concern is that these fiscal reductions have come when society is more dependent on the public services offered by the library. However, when the library has established strong community ties, budget restraints become much less tolerated. Many county libraries report patrons protesting such reductions in funds (WBOC16, 2013; Vande, 2014; Wescott, 2012) and the one common denominator in each case is that these libraries are supported by informed patrons, demonstrating that the strong emotional attachment to the library coupled with a thorough understanding of exactly what it does results in a population that will fight budget cuts.

An example of this step in the ACE model is the Charlotte Mecklenburg Library in North Carolina. In 2010 Charlotte Mecklenburg faced a potential $2 million deficit and the closure of half of the 24 branches within the system. Communications specialist Jennifer Daniels was sending live Twitter messages from a board meeting asserting her astonishment at the reductions proposed, and an already-engaged community in social media was listening. From that one post, word spread via Facebook, and more than $400,000 was eventually raised in the local neighborhoods, resulting in the closure of only four of the proposed 12 libraries (Dankowski, 2013, p. 40).

Even with strong support, there may be a need for additional revenue streams to fill gaps created by less government funding, and this also falls back to a strong relationship between the library and its community achieved through engagement, as seen in step five of the model. Many libraries have now enacted programs and services that not only bring more people into the library, but also enhance fund raising. Additional revenue is achieved in many ways--renting out rooms for weddings, hosting cafes, offering unique classes, selling used books that have been donated by patrons, even running gift shops. Surveys of libraries show that revenues can increase as a result of fee-based services (Dempsey, 2010, p. 22).

This is not to suggest that the library become a for-profit center. A desire to bring in extra funds should not be viewed as changing the purpose of the institution, but rather one way to mitigate budget reductions. The result is a library with an adequate budget, well connected to its community, and in a better position to serve.

With the more substantial budgets that can result from step five in the model, the library will be better positioned to offer longer operating hours, hire more professionals, enhance technology (such as 3-D printers and robotics), and improve collections.

Community members now have more reasons to visit their library. Which brings us to the sixth and final step of the model: new services and programming. These also increase communication between the library and its community. Thus the cycle continues as illustrated in Figure 1.

Having cycled through the model, the library has employed a wide variety of tools to connect with its public, resulting in a better-informed community that is more supportive of the library. This can positively impact budgets, allowing libraries to expand services, thereby continuing to increase patron affinity for the institution.

On a final note, the model presented here is titled "adaptable" because other sectors can benefit by employing similar outreach. A hospital can connect better with its local community through public engagement. Schools can clearly benefit when opening the doors to a two-way conversation. Even for-profit entities would do well to work hand in hand with the public. With minor tweaks and more appropriate verbiage, the ACE paradigm could be revised for other institutions.

REFERENCES

1. American Library Association. (2013). *ALA Policy Manual.* Retrieved from http://www.ala.org/aboutala/governance /policymanual/updatedpolicymanual/secti on1/1mission.

2. American Library Association. (n.d.). About LTC. Retrieved from Libraries Transform website: http://www.ala.org/ transforminglibraries/libraries-transforming- communities/ about-ltc.

3. Cahill, K. (2011). Going social at Vancouver Public Library: What the virtual branch did next. *Program: Electronic Library and Information Systems, 45*(3), 259-278.

4. Dankowski, T. (2013). How libraries are using social media. *American Libraries, 44*(5), 38-41.

5. Dempsey, B. (2010). For love or money. *Library Journal, 135*(15), 20-23.

6. Dumais, S., Cutrell, E. & Chen, H. (2001). Optimizing search by showing results in context. In M. Beaudouin-Lafon & R. J. K. Jacobs (Eds.), *Proceedings of ACM CHI 2001 Human Factors in Computing Systems Conference)* (pp. 277-284). ACM Press.

7. Eckerle, C. (2013, January 31). Social media marketing: How New York Public Library increased card sign-ups by 35% [Blog post]. Retrieved from http://sherpablog. marketingsherpa.com/online-marketing/nypl-social-media-marketing/

8. Finley, M. F. (2000, April 10). Education leads to better understanding. *Spokesman- Review* (Spokane, WA).

9. Henderson, E., & Lonergan, J.. (2011). *Majority of states report decline in support for library services*. Research Brief series, no. 3 (IMLS-2011-RB-03). Washington, DC: Institute of Museum and Library Services.

10. Museums, Libraries & Archives. (2011). *Community engagement in public libraries: An evaluation of the Big Lottery Fund's Community Libraries Programme*. Birmingham, UK.

11. Vande, B. M. (2014, May 21). Union protest at Grand Rapids Public Library: Will patrons suffer from job cuts? *The Grand Rapids Press: Web Edition Articles (MI)*.

12. WBOC16. (2013, February 20). Library patrons fight against budget cuts. Retrieved from http://www.wboc.com/story /12463766/library-patrons-fight-against-budget- cuts.

13. Wescott, S. (2012). A library fights budget cuts with social-media campaign. *The Chronicle of Philanthropy*. Retrieved from https://philanthropy.com/article/How-a- Library-Used/156961

14. Wooden, R. A. (2006, Winter). The future of public libraries in an Internet age. *National Civic Review*, *95*(4), 3-7

Chapter 13

User-Experience Design and Library Spaces: A Pathway to Innovation?

John A. McArthur   Valerie Johnson Graham

Queens University of Charlotte

ABSTRACT

Libraries have responded to the rapid change in communication and information technology by developing an understanding of how their clienteles perceive and plan to use libraries in the 21st century. This article positions user-experience design and specifically Don Norman's ideas about behavioral, cognitive, and reflective responses of consumers to products as a pathway for libraries to innovate through spatial design and behavioral practices. After a brief introduction to experience design, this study connects Norman's design levels to emerging trends and innovations in library spaces.

Many libraries are seeking ways to ensure that their physical spaces have continued meaning and value. Traditionally, libraries have been viewed as content providers and storehouses for reference materials and historical information (Dillon, 2008). But with advances in digital technology and the ubiquitous availability of information, the notion of the experience offered by libraries is changing (Jui, 1993; Jiao & Onwuegbuzie, 2004; Bertot, McClure, & Jaeger, 2008; Holmberg,

Huvila, Kronqvist-Berg, & Widén-Wulff, 2009; Anttiroiko & Savolainen, 2011). Libraries are measuring the value and impact of digital content and pivoting accordingly to meet the needs of patrons.

This trend toward digital consumption has been matched with a call for libraries to redefine the purpose and uses of their physical spaces (Hanke, Slaughter, & Watkins, 2012; Gray & Copeland, 2012). User-experience design principles have emerged as one avenue for considering the role of physical space in the experience of libraries (Schmidt 2010, 2012a, 2012b). User-experience design (UXD) positions the role of a product as interactive with the user (Shedroff, 2001). Williams (2007) provides an excellent analysis of this emerging field, concluding that an experience design perspective strives for "designs that stimulate and satisfy our intellects; that please our emotions; and that engage our senses all while helping us to achieve an instrumental goal" (p. 1). Williams' description of UXD highlights the importance of both the effectiveness of design and the interactions between a designed object and its users.

USER-EXPERIENCE DESIGN AND THE SPACE OF LIBRARIES

This article furthers the optimistic approach to the future of libraries (as noted by Giesecke, 2011; Stoffle & Cuiller, 2011) by investigating the ways that user-experience design theories can address the changing roles of contemporary libraries. Specifically, this article will describe how libraries have begun the process of investing in this process of innovation by applying Norman's (2004) user-experience design framework to their physical space. Donald A. Norman, co-founder of the Nielsen Norman Group and an academic in cognitive science, design, and usability engineering, offers a design model in his book, *Emotional Design: Why we love (or hate) everyday things* (2004). He describes three levels of emotional processing: visceral, behavioral, and reflective. He suggests that each of the three levels plays a different role "in the total functioning of people" and each requires a different aspect or style of design (2004, p. 21).

Although Norman limits the three levels of design to products, we contend that these levels of design are also applicable to physical spaces. Norman even states that, "perhaps more significant (than products), however, is our attachment to places: favorite corners of our homes, favorite locations, favorite views" (p. 48). For many, libraries are also favorite places and are associated with memories and feelings. The attachments people associate with spaces was the impetus for applying Norman's theory to the experience of libraries and other spaces. By using Norman's visceral, behavioral, and reflective design principles to evaluate and repurpose their physical spaces, librarians will have an opportunity to discover unmet and unarticulated needs within their communities as well as uncover possible design approaches to meet those needs.

VISCERAL DESIGN IN LIBRARIES

Norman explains the visceral level as the initial response, the gut reaction, to available sensory information. Norman list things such as warmth, comfortable lighting, harmonious music and other sounds, symmetrical objects, smiling faces, or sweet tastes and smells, as things that tend to give rise to positive affect. However, he identifies things such as darkness, extreme hot or cold, crowds of people, sharp objects, harsh abrupt sounds, or bitter tastes as things that tend to inspire a negative affect (2004, p. 29-30). The visceral is the level on which first impressions are formed.

Principles of visceral design are innate for humans. They are based on "initial reactions" and "immediate emotional impact" (Norman, 2004, p. 68-69). Because physical features such as look, feel, and sound dominate, a user's response to a product at this level will always inspire love or hate, attractiveness or unattractiveness (Norman, 2004, p. 67).

The best visceral design will ignite an "I want it" reaction without the person first asking what the product does or how much it costs (Norman, 2004, p. 68-69).

Queens Library in Far Rockaway, New York, undertook one successful approach using principles of visceral design. Queens Library is a heavily used library located in an area with high unemployment and a lot of gang activity (Hinkle, 2010). Teens often hung out in the library, which caused disruption for the other patrons. To overcome this problem, the Far Rockaway library received grants to create space called the Queens Library for Teens, an extension located a block away from the main library. This space was designed solely for teens (Bolan and Nelson, 2008) based on focus group research targeting the age group. The research revealed attention to physical design of spaces (e.g., color, furniture, and lighting), as well as furnishings present (e.g., music booths, video game equipment, flat-screen TVs).

Visceral design surrounds users' initial reactions and first impressions. The Queens Library for Teens created an inviting space for its target user by employing visceral design principles while redesigning the space. By having brightly painted walls and furnishings and providing teen-approved gaming and computing technology, the library is able to maintain positive initial reactions and first impressions from its teen users.

Moreover, the services provided by the library reinforce the design. Some distinguishing features of the library include specialized programming for open mics or Wii gaming, lack of a ban on mobile phone use and food, a vocal recording booth, and 40 Internet-capable computer stations. The library provides magazines and online resources for the teens but doesn't circulate books (Hinkle, 2010). Through color, furnishings, and signage, these spatial alterations to the human environment create a visceral reaction for users that is inviting and tailored to a specific subgroup of patrons.

London's Tower Hamlets Libraries discovered that less than 20 percent of their library patrons actually set foot in the library (Sudjic, 2004). At Tower Hamlets, the staff redesigned the physical environment to alter patrons' impressions of the library. Instead of the typical layout of a library, spaces called "Idea Stores" were created. They incorporated retail style design principles and are placed in high-traffic shopping areas of the city. Sudjic (2004) describes the first impression of one of the libraries as a "multicolored ribbon of glass built on an abandoned roof garden on top of a single-story mall of shops, making a site out of almost nothing" (para. 8). Even the job descriptions reflect this design. One of the positions in an Idea Store is the "window-dresser," a person in charge of creating window displays to highlight the store (Patterson, 2001).

Norman's concept of good visceral design for products suggests that users simply realize they like the design. In the case of the Idea Store, people passing by may not be concerned with what the space actually is, but they know they want to go inside.

Capitalizing on the visceral response of the user can be a valuable first step for libraries to entice users to see the library with fresh eyes.

This knowledge of visceral design and the examples above lead to the following *opportunities for libraries to innovate in the area of visceral design:*

- Creating visually stunning entry points. Library entryways might use large, single graphic elements to catch user's attention before and upon entry.

- Inviting the outside in. Often achieved through glass walls or large, clear and unobstructed windows, but this could also include live video feeds or other displays on the exterior of a building that give clues to what is happening inside.

- Considering intentional, and varied, auditory cues. Spaces in libraries with different functions might incorporate varied sounds. A low, white noise designed for a children's area might include different sounds than one in a computer center or a location intended for quiet study.

- Exploring options for tactile and olfactory cues. The physical feel of a space (door handles, flooring, hard and soft surfaces) as well as the smells of a space contribute to users' visceral impressions, even though they might not often realize its importance.

These examples are positioned as a starting point—an initial brainstorm to spark local ideas—for libraries seeking to innovate in the area of visceral design. As more libraries investigate their own physical environments, opportunities to harness patrons' visceral responses can be tailored to local audiences for significant impact.

BEHAVIORAL DESIGN IN LIBRARIES

The behavioral level of design is most notable during the interaction between product and user and encompasses the emotions a person has while using a product.

Appearance and rationale matter less in this level, and a product's performance during use becomes increasingly important. This study argues that, out of the three levels of design, libraries have been the most successful in implementing behavioral design principles to their spaces. Norman (2004) investigates four principles for good behavioral design: function, understandability, usability, and physical feel (2004, p. 70), which will be used to frame examples herein.

FUNCTION

Function represents the product's overall purpose and the need that it fulfills for the user. Good design would therefore meet the purposes of the user. Norman (2004) uses the example of a teapot with the spout and handle on the same side. A user might be drawn to the beauty of this teapot, but its function might not meet the needs of someone wanting to pour tea.

Function can both apply to the things users know they need in a space, but also to services which users were never aware they wanted or needed. Schmidt (2012a) gives examples of libraries across the United States doing just that. For instance, Baltimarket (http://www.baltimarket.org) is a virtual supermarket that allows customers to order groceries online and then pick them up at their local library. H.O.M.E. Page Café (https://projecthome.org/café), part of the Free Library of Philadelphia, includes the goal of providing jobs to formerly homeless people in its central purpose. Schmidt (2012b) points out these new services are "based on specific needs" and they have placed libraries "at the center of satisfying those needs" (para. 12). In these two examples, the physical space of the library has been aligned with functions not normally associated with libraries. From a behavioral design perspective, these innovations offer new ways for libraries to connect with their communities.

USABILITY

On the issue of usability, Norman (2004) argues that a product may function as it is supposed to and be easy to understand yet it may remain unusable to some people. For example, the design of a library kiosk might include issues of access to the kiosk for people of differing heights or physical abilities, language settings on the display, or the size and shape of items a user is attempting to check out. According to Norman (2004), successful usability design is the most difficult to achieve. The only thing that matters for usability is ability of a user to successfully and competently utilize the design. For example, usability studies have been used to measure compliance with the Americans with Disabilities Act and to correspond with a growing focus on universal design among building planners.

Usability might further be explored in the connections between physical spaces and e- branches, those divisions of libraries with staff allocated for online interaction.

Rochester Hills Public Library (RHPL) developed an e-branch so efficient that some patrons who are well-versed with technology suggested they had no need to visit the library. The e-branch provides everything from e-books to resources needed for research and library assistance (Lind Hage, 2012). This high degree of successful usability devalues physical space. The momentum toward online resources is indicative of the collective appreciation for and attention to usability by libraries for many years.

But this momentum suggests that one goal for behavioral design of library spaces might be to consider not only the usability of digital resources, but also the usability of physical spaces as redesigned for new uses.

The Pew Research Center's 2013 study (Zickuhr, Rainie, Purcell, & Duggan, 2013) on libraries in communities noted that 48% of the American population visited a public library in 2013, down from 53% in 2012. Yet, 72% of American households had at least one member visit a library in the same year. Even though the same report notes an increase in visits to web-based library resources, the report notes that users value having a quiet, safe place almost as such as having access to books and media.

Patrons visit libraries not only for access to content but also for access to usable space (Zickuhr, et al., 2013). This study underscores the need for libraries to consider not only the usability of their resources, but also the usability of physical spaces for re-designed purposes.

UNDERSTANDABILITY

Norman (2004) suggests that the concept behind the design should be relatable to both designer and user. As users approach the product, the design communicates with the user and explains itself. In an ideal world, the designer's idea and the user's expectations would be identical, resulting in the user understanding and using the product in the way intended by the designer (Norman, 2004, p. 75). For instance, referring to the kiosk example, when the user places an item (correctly) in the scanner, the item number usually registers on the kiosk. If the item was placed incorrectly in the scanner and the user cannot discern correct placement, frustration or negative emotions may set in, due in part to a lack of understanding of how the machine works. This, Norman argues, would be an issue of design. The product, in this case the kiosk, was not designed to be understandable to the user.

Barlow and Morris (2007) examined library understandability from the perspective of new users by assessing their experiences at traditional libraries (those which have not been refurbished in the last five years) and at new libraries (those built between 2004 and 2006). The authors used a variety of methods to examine usability from the perspective of a new user: walk through audits, associated task analyses, interviews, and questionnaires. With these various methods, Barlow and Morris concluded the study with a "best practice guide" to aid in future library design. While Barlow and Morris (2007) allude to all three of levels of design discussed by Norman, they reinforce behavioral design by noting the importance of understandability. Understandability in libraries often derives from signage and a user's wayfinding experience. By having a positive experience with signs, users are likely to be able to navigate the library more efficiently. These directories, wall signs, and hanging signs point patrons not only to various circulation departments and customer service desks, but also to restrooms, seating and dining areas, and group meeting spaces. In addition, wayfinding might also be a consideration for navigation of library holdings. Patrons and librarians often navigate the shelves of books and media in search of specific titles, organized in a particular wayfinding model. Understanding a library's encoding process for library holdings is a key understandability issue.

One further consideration around understandability may even be related to the terminology of the word *library*. Patterson (2001) and Sudjic (2004) both discuss libraries that disassociated their work from the term library. In the 2013 Pew Study (Zickuhr, et al., 2013) mentioned above, only 23% of library patrons report knowing about most of the services offered by libraries. The vast majority of patrons reported knowing only some or little about available services (Zickuhr, et al., 2013). Perhaps this finding adds merit to the assertion that the general public continues to view contemporary libraries as the storehouses of the past despite the new services being offered. While little academic research has been conducted on the issue of library naming, anecdotal information from library systems may be able to demonstrate this pattern of name shifting to accompany a shift in roles. This remains an issue of understandability.

PHYSICAL FEEL

In the arena of physical feel, Norman (2004) claims that the touch and feel of a product is correlated to the appreciation one may have for it (p. 79). Norman (2004) argues,

Far too many high-technology creations have moved from real physical controls and products to ones that reside on computer screens, to be operated by touching the screen or manipulating a mouse. All the pleasure of manipulating a physical object is gone and, with it, a sense of control. Physical feel matters. We are, after all, biological creatures, with physical bodies, arms, and legs. (p. 79)

In the kiosk example, this might include the feel of the scanning area, the smoothness of the display screen, the temperature of the machine areas, or the ease of motion associated with scanning an item.

Dewey (2008) discusses the University of Tennessee Libraries' Commons. Phase Three of The Commons Plan is to maintain the physical space of the library as an "inspiring place" by adding comfortable yet flexible furniture for individual and group work. This is a fairly common refrain for libraries, but the point is clear: By improving the physical feel of library spaces, users' feelings of pleasure and comfort will increase (for further reading on physical feel and product design, see Coates, 2003; Jordan, 2000).

Knowledge of the four aspects of behavioral design and the examples above lead to the following opportunities for libraries to innovate in the area of behavioral design:

- Ensuring that form follows function. The patchwork of important uses filled by a single library space requires multiple dedicated spaces to support multiple uses. Whereas the local library and its patrons best determine these, the continuous assessment of space and its uses allows a library to pivot with the needs of its patrons.

- Devising digital-physical intersections for library spaces. These might include mobile apps that function as signage, indoor mapping applications that lead patrons to resources, or primer videos for decoding the placement of library holdings.

- Emphasizing naming in the library space. The naming and marketing of services common to the local library enables users to learn about and experience them. The first naming challenge for libraries appears to be the gap in user knowledge about available services. Naming and directional cues can address behavioral issues surrounding access to and use of these services in the library's physical environment.

- Attending to issues of physical feel. Like all public spaces, the furnishings in libraries deteriorate over time. Libraries might build a rotation of furnishing improvements that follows the categories of uses present in the space, allowing the cycle of refurbishment to follow the patterns of use in the space. Again, these examples are positioned as a starting point – an initial brainstorm to spark localized ideas – for libraries seeking to innovate in the area of behavioral design.

REFLECTIVE DESIGN IN LIBRARIES

In Norman's (2004) reflective level, culture, experiences, and memories come into play in the cognitive work of interpretation, understanding, and reasoning. Therefore, reflective design should be about "long term relations" and "the feelings of satisfaction produced by owning, displaying, and using a product" (Norman, 2004, p. 38). Norman claims that a person's self-identity is found at the reflective level. Interaction between a person's identity and the product are demonstrated by either a sense of pride or shame accompanying ownership or use.

Properties of reflective design are twofold: memory and self-image. In memory, customers will reflect back on the appeal and experience with the product. For Norman (2004), "pleasant reflective memory can overcome any prior negative experiences" (p. 88), meaning that if a user thinks back positively about a library program, that memory may outweigh the fact that she had to wait in line to enter. In terms of self-image, Norman argues that the design of any product is a reflection of who we are, and our use of a product signifies our association with the product's image.

For example, phase two of Dewey's (2008) exploration of the University of Tennessee's Libraries' Commons demonstrates the role of the library as a social/cultural space. Just as assumptions are made about a person by what they wear, the same can be said for the spaces or places they frequent. Dewey describes the way a formally built Starbucks in the space has become a popular place for students and faculty to meet and socialize, and a Cyber café has workstations used for emailing and social networking. By signifying their use as a social/cultural space, Dewey reiterates the concept that, for the students and faculty using the space, their physical presence makes a statement about their social and cultural characteristics.

Reflective design in libraries also would emphasize the lasting impact of the library in the mind of the user and the user's public identification with the library. The Sno-Isle Libraries in Washington State faced a challenge in which users assessed their library visits as positive experiences, but those experiences failed to translate to advocacy and donor support indicative of identification with the library. To build a better relationship with their community and spread advocacy, the Sno-Isle Libraries developed a "Community Ambassador" program (Telford, 2012). The library recruited individuals who had positive experiences and trained them to tell others about their experiences. The reflective design principles applied could be useful in creating the support and long-term relationships the Sno-Isle libraries are seeking.

Many libraries are using social media to connect users, positive experiences, *and promotion.* The Charlotte-Mecklenburg Libraries in Charlotte, NC are not only filled with positive messages about how the library engages the community, but also invitations for users to tag the library in their Facebook, Twitter, and Instagram updates (https://cmlibrary.org/our-brand-promise). This social media engagement connects users and libraries in the arena of reflective design through immediate identification with the brand on public platforms. This connection between digital and physical space will continue to be a fruitful intersection for library spaces in reflective design.

In a non-traditional yet memorable approach to gaining community support, the Mayfield Library in Dalkeith, Midlothian, Scotland held a pole dancing class as part of "Love Your Library Day" (Pfeiffer, 2013). The class served as a way for the library to reach more patrons with the intent of increasing book circulation and also promoting other services offered by the library Although a rather unconventional approach to inspiring reflective thinking, this library is engaged in unique, but also reflective, design. When applying reflective design principles to spaces, this study argues that not only must the space be viewed in a positive manner, but that view must also lead to long-term, supporting relationships.

Considering reflective design and the examples above, the following are opportunities for libraries to innovate in the area of reflective design:

- Capitalizing on emerging social media trends among target populations. In 2014, the opportunities to share selfies, engage friends on SnapChat, pin items on Pinterest, and tag geo-located spaces on Facebook were novel ideas in the social media space. Posting in the social media space, in this context, provides a moment (however brief) of reflection on interaction with the library.

- Inviting real-time engagement in the library space. A monthly contest for selfies could incorporate pictures in the library with favorite books, favorite seats, favorite librarians, or help to educate other patrons about available services. Libraries might use simple applications to have them appear in real time on digital boards inside the library.

- Training library ambassadors. Giving people language to discuss their experiences with libraries can help a community learn how to support its local library.

- Archiving and sharing stories. Retaining the oral history surrounding a local library allows community members to consider their relationship to libraries and to publicly show the value of libraries. Video recordings of these can be stitched together with any number of digital applications.

- Hosting opportunities for reflection. The physical space of a library can and should be a site of reflection for patrons. Future library programs might include interviews with local patrons about the role of libraries in communities.

As stated previously, these examples are positioned as a starting point – an initial brainstorm to spark local ideas – for libraries seeking to innovate in the area of reflective design.

Concluding Thoughts

In discussing the University of Tennessee's Libraries, Dewey (2008) emphasizes the importance of physical space, stating, "If we, as educators, truly believe that our expensive collections and numerous services are relevant and vital to the intellectual development of students and the process of creating knowledge we must do everything we can to be in the same space" (p. 86). For libraries to invest in our communities, we must not only face outward into community events, but we must also create physical spaces which compel people to enter. User-experience design has the ability to aid libraries through innovation in physical space.

The many success stories presented herein suggest that the role of the physical space of libraries has drastically changed and that library administrators recognize these changes. By assessing these examples through the lens of user-experience design, this study has argued for the important role of design theory in future library planning, particularly in considerations of the physical environment. Norman's (2004) framework for product design has significant value for assessing physical space in libraries as well as digital installations, e-branches, and website-based interactions. Future studies in this area are needed. First, libraries could seek to assess the visceral, behavioral, and reflective designs present in library spaces through focus groups and other formal interviews and observed interactions with patrons. Such studies will allow libraries to gain perspective about how their physical spaces are used and perceived. Then, the pivot point for designing physical space is the behavioral connection between function and usability. This shift in function coupled with a library's continued investment in usability has numerous implications for library spaces ready for research: the role of librarians as hosts in the physical space, the time spent by patrons in the facility, the optimal circulation goals for printed materials, and the behavioral norms that users associate with the term library. Moreover, this article also positions the digital-physical interaction of libraries and patrons as an avenue for future research. As patrons engage in the processes of memory and self-image in reflective design, increased brand loyalty and library support may ensue.

Although they were outside the scope of this article, libraries' purely digital points of connection to users may also be a fertile ground for study using user-experience design principles. Dillon (2008) describes E-branches, library websites, and information portals as viable vehicles for interaction between a library and its patrons. These digital sites may well become spaces in which users create meaningful perceptions about the availability of library resources and the functions of a library in the community.

Administrators, users, and patrons of libraries each bring a personal bias to the understanding of "library" as a space and place in a community. Libraries around the world are faced with the challenge that people think of the library in the traditional sense of an information storehouse rather than envisioning it as a more contemporary and networked information hub. Studies and research into reflective design would benefit the community of library administrators. As this article has already addressed, Patterson (2001) and Sudjic (2004) discuss the disassociation of the term "library" from some of the services being developed. However, this article, in turn, calls for libraries to revisit the user experience in libraries as a design-based solution. The principles espoused by Norman (2004) articulated herein can give libraries a framework for self-assessment and external review, as well as a path forward for innovating in the built spaces they occupy. At the very least, these principles offer library staff members a path for considering the future of their own individual libraries through the experiences of the user.

REFERENCES

1. Anttiroiko, A., & Savolainen, R. (2011). Towards library 2.0: The adoption of Web 2.0 technologies in public libraries. *Libri: International Journal of Libraries & Information Services, 61*(2), 87-99.

2. Barlow, A., & Morris, A. (2007). Usability of public libraries: perceptions and experiences of new users. *IFLA Conference Proceedings,* 1-21.

3. Bertot, J. C., McClure, C. R., & Jaeger, P. T. (2008). The impacts of free public Internet access on public library patrons and communities. *Library Quarterly, 78,* 285- 301.

4. Bolan, K., & Nelson, J. (2008, September 18). The Far Rockaway teen library provides offsite space to be, well, a teen. *Library Journal.* Retrieved from http://lj.libraryjournal.com/2008/09/ljarchives/just-for-them/

5. Coates, D. (2003). *Watches Tell More Than Time: Product design, information, and the quest for elegance.* New York: McGraw-Hill.

6. Dewey, B. I. (2008). Social, intellectual, and cultural spaces: creating compelling library environments for the digital age. *Journal of Library Administration, 48*(1), 85-94.

7. Dillon, D. (2008). A world infinite and accessible: Digital ubiquity, the adaptable library, and the end of information. *Journal of Library Administration, 48*(1), 69-83.

8. Gisecke, J. (2011). Finding the right metaphor: Restructuring, realigning, and repackaging today's research libraries. *Journal of Library Administration, 51*(1), 54-65.

9. Gray, D. J., & Copeland, A. J. (2012). E-Book versus print. *Reference & User Services Quarterly, 51*(4), 334-339.

10. Hanke, E., Slaughter, A., & Watkins, C. (2012). E is for elephant… and E-books. *ILA Reporter, 30*(2), 4-8.

11. Hinkle, S. (2010, November 7). Queens & teens [Blog post]. Retrieved from http://libraryhearted.blogspot.com/2010/11/queens-teens.html

12. Holmberg, K., Huvila, I., Kronqvist-Berg, M., & Widén-Wulff, G. (2009). What is library 2.0?. *Journal of Documentation, 65*(4), 668-681.

13. Jiao, Q. G., & Onwuegbuzie, A. J. (2004). The impact of information technology on library anxiety: The role of computer attitudes. *Information Technology & Libraries, 23*(4), 138-144.

14. Jordan, P. A. (2000). *Designing Pleasurable Products.* London: Taylor & Francis, Inc.

15. Jui, D. (1993). Technology's impact on library operations. [Washington, D.C.]: Distributed by ERIC Clearinghouse, http://www.eric.ed.gov/contentdelivery/servlet/ERICServlet?accno=ED389333

16. Lind Hage, C. (2012). Will a new E-Branch increase library usage?. *Public Libraries, 51*(4), 12-13.

17. Norman, D. A. (2004). *Emotional design: Why we love (or hate) everyday things.* New York, NY: Basic Books.

18. Patterson, T. (2001, May 1). "Idea stores": London's new libraries: An experiment in the blighted East End redeploys and revamps library service. *Library Journal, 126*(8), 48-49.

19. Pfeiffer, E. (2013, February 4). Library offers free pole dancing class to draw visitors. Retrieved from http://news.yahoo.com/blogs/sideshow/local-library-offers-free- pole-dancing-class-draw-213912777.html

20. Schmidt, A. (2010). The user experience. *Library Journal, 135*(1), 28-29. Schmidt, A. (2012a). Stepping out of the library. *Library Journal, 137*(4), 26. Schmidt, A. (2012b). Experience better than ebooks. *Library Journal, 137*(8), 17.

21. Stoffle, C.A., & Cuillier, C. (2011). From surviving to thriving. *Journal of Library Administration, 51*(1), 130-155.

22. Sudjic, D. (2004, July 10). When is a library not a library? When it's an 'idea store'. *The Guardian.* Retrieved from http://www.guardian.co.uk/artanddesign/2004/jul/11/art2

23. Telford, C. (2012). *Community ambassador: Sno-Isle Libraries.* Retrieved from Urban Libraries Council website: http://www.urbanlibraries.org/community-ambassador-innovation- 183.php?page_id=40

24. Williams, S. D. (2007). User experience design for technical communication: Expanding our notions of quality information design. *Proceedings of the Annual Meeting of the IEEE Professional Communication Society.*

25. Zickuhr, K, Rainie, L., Purcell, K., & Duggan, M. (2013). *How Americans value public llbraries in their communities.* Retrieved from Pew Internet website:http://libraries. pewinternet.org/2013/12/11/libraries-in-communities/

Chapter 14

Academic Education in Library and Information Management in Bulgaria

Rositsa Krasteva[1], Tereza Trencheva[2], Sabina Eftimova[2], Tania Todorova[2]

[1]Computer Science Department, State University of Library Studies and Information Technologies, Sofia, Bulgaria.
[2]Library Management Department, State University of Library Studies and Information Technologies, Sofia, Bulgaria.

ABSTRACT

The purpose of this publication is to present the contemporary aspects of training educational and qualification degree "Bachelor" of Specialty "Library and Information Management" of the Library Management Department at the State University of Library Studies and Information Technologies (SULSIT) in Sofia, Bulgaria.In view of specificity and completeness of the presented information in this publication there is a limit, which refers to the training only in educational and qualification degree "Bachelor". The following methods are used: a study of the curriculums of many universities worldwide teaching in this or a related specialty; comparative analysis; synthesis of the obtained information. Accent is put on the disciplines Intellectual Property, Standardization in Library

Activities, Quality Management in Library and Information Activities, Library Psychology and Bibliotherapy. The research draws attention to some aspects insufficiently covered in the curriculum of the programs preparing future highly knowledgeable, trained library and information managers, and offers some solutions, based on our experience in the State University of Library Studies and Information Technologies, to the attention of the LIS academic and professional community.

Keywords: Library Management, Information Management, Higher Education, Bachelor's Degree, SULSIT, Bulgaria

1. INTRODUCTION

Socio-economic and political transformations in the Bulgarian reality necessitated a change in the functions of libraries and enhanced their role in the process of accession to the European and World educational, cultural and economic space. Bulgaria is member of the EU since January 1st, 2007. The libraries in accordance with the principle of wide and equal public access to library and information resources, and as a democratic social institutions contribute to social stability, preservation and development of the spiritual and scientific potential of society. The contemporary challenges facing the libraries, suggest the application of adequate library management. It evolves against the background of changing the overall management paradigm—a process characteristic of global management. The outdated management approaches to the concept of "Rational Management" shall be replaced by the Marketing Paradigm, characterized by adaptability and flexibility of management. This undoubtedly imposes new requirements, including management knowledge and skills to the competencies of young specialists, finishing their high degree in specialties like Library and Information Science.

Evaluating LIS Education in Europe, the experts of the European LIS Curriculum Project conclude that in Europe, almost all library and information (LIS) programs in higher education have been developed and are offered within the context of a nation-state [1] . Several authors reported the need for internationalization of library education in Europe in accordance to achieve greater results in student and teacher mobility [2] -[4] . The situation is summarized in LIS Education in Europe: "The Structure and contents of LIS courses vary very much between the different types of LIS education providers in Europe, which include many fairly small academic environments. The apparent disparate nature of LIS educational programmes in Europe constitutes a barrier to increased cooperation in the field. There is a marked need for joint discussions of the structure and contents of LIS school curricula and for identifying and discussing possible common curricular elements both for the purpose of enhancing the quality of individual LIS educational programmes and for the sake of increased collaboration between European LIS school programmes" [5] . Based on these findings, in 2005 started the development of an international project LIS education in Europe, coordinated by the Royal School of Library and Information Science, Denmark. Main goals of the project are:

- To create better opportunities for European mobility of students and teachers;

- To increase the scale of mobility and inter-institutional cooperation;

- To develop a common conceptual framework to define the core of mandatory elements within the LIS curriculum as a basis for increasing the mobility of students and teachers and the acceleration of the Bologna process;

- To work for greater flexibility, transparency and comparability of curricula.

The results of this project were published in the European Curriculum Reflections on Library and Information Science Education [6] . In a separate section was provided information related to the training of library management. Focus is on disciplines such as Marketing Information Services; Communication Skills and Negotiations; Intellectual Property and Information Law; Quality Management and etc.

Taking into account the achievements of our colleagues in Europe, in this article we would like to share our experiences, best practices and suggestions on training in the undergraduate program "Library and Information Management" at the State University of Library Studies and Information Technologies in Sofia, Bulgaria. In response to the social and professional needs, in 2008 in the structure of the Faculty of Librarianship and Cultural Heritage (FLCH) in SULSIT is created new Department of Library Management. Founder and first head of the department was Prof. DSc Ivanka Yankova (from 2011 she is a Dean of FLCH). Since June 2011 up to now— Assoc. Prof. PhD. Tania Todorova is a Chair Holder of Library Management Department. The main task of the Library Management Department is the preparation of bachelors, masters and doctoral students in the field of library and information management.

2. GOAL, RESEARCH TASKS AND METHODOLOGY OF THE STUDY

The purpose of the study is to present the contemporary aspects of training in educational and qualification degree "Bachelor" of the Specialty "Library and Information Management" of the Library Management Department at the State University of Library Studies and Information Technologies (SULSIT) in Sofia, Bulgaria and to make comparison with the curricula of some of the biggest universities worldwide, which are accredited in LIS education. In view of specificity and completeness of the presented information in this publication there is a limit, which refers to the training only in educational and qualification degree "Bachelor". To achieve the objective are set out the following major research tasks, outlining the methodology of the study: 1) Presentation of the Library Management Department in SULSIT (Bulgaria) and its activity; 2) Exploration, presentation and comparative analysis of the curriculums of many universities worldwide teaching in this or a related specialty; 3) Conclusions and recommendations. The methodology for achieving the objective of the study and solving the set research tasks include the following specific methods: method of study and content analysis, comparative analysis; synthesis of the obtained information. Accent is put on the disciplines Intellectual Property, Standardization in Library Activities, Quality Management in Library and Information Activities, Library Psychology and Bibliotherapy. The research draws attention to some aspects insufficiently covered in the curriculum of the programs preparing future highly knowledgeable, trained library and information managers, and offers some solutions, based on our experience in the State University of Library Studies and Information Technologies, to the attention of the LIS academic and professional community.

3. INTERNATIONAL COOPERATION OF THE LIBRARY MANAGEMENT DEPARTMENT

The creation and development of the Library Management Department and proposed programs—Bachelor Program "Library and Information Management" and Master Program "Library, Information and Cultural Management" are influenced by the principles and guidelines of the Bologna Declaration and the study and application of leading experience of European and foreign universities.

In the period 2009-2013, the Library Management Department was a partner in two Erasmus Intensive Programmes with the participation of European universities. The Project Das Grimm-Zentrum-(k)ein Bibliotheksmärchen) (2009-2011) with coordinator Humboldt University in Berlin (Germany) created a unique opportunity for cooperation with University of Vienna (Austria), Vilnius University (Lithuania), Masaryk University Brno (Czech Republic) [7] . In the period 2011-2013 Library Management Department at SULSIT was coordinator of Erasmus Intensive Programme "Library, Information and Cultural Heritage Management—Academic Summer School" [8] .

Project Manager is Tania Todorova—Head of Library Management Department. Our partners are: Hacettepe University in Ankara (Turkey), University of Zagreb (Croatia), University Paris Descartes (France) and University of Szeget (Hungary). The mission of this project is through using a rich methodological toolkit to implement a modern educational process aimed at implementation of interdisciplinary knowledge and skills relevant to the new requirements in the career development of students in library, information and cultural sector and the policy response to higher education and the EU initiative on "New Skills for New Jobs". The main topics in IP LibCMASSare: Library, Information and Cultural Management; Information Literacy; Preservation and access to Cultural Heritage; Digital libraries; Intellectual Property; Information brokerage; Information technologies in libraries, archives, museums and other cultural institutions [9] [10] .

In the process of creating and updating the curriculum of the Bachelor Program "Library and Information Management" important was the impact of active participation of professors and students from the Library Management Department in the short-term mobility under the Erasmus Intensive Programmes. On Figure 1 is shown the dynamics of the participation of professors. On Figure 2 is presented the students participation. It is necessary to highlight that in 2013 was implemented successfully 1 student mobility of an Erasmus Summer Semester in HochschulefuerMedien in Stuttgard, Germany.

These projects were unique opportunity for establishment a stable international network in higher education in Library and Information Sciences, Computer Sciences and Cultural Heritage Sciences and to promote cooperation between academic education and practice— library, information and cultural sector.

On 20th November 2011 the Memorandum for establishment of the UNESCO Chair' ICT in Library Studies, Education and Cultural Heritage' was signed between Irina Bokova, the General Director of UNESCO and Prof. DSc Stoyan Denchev, the Rector of the State University of Library Studies and Information Technologies. Many colleagues from Library

Management Department collaborate actively in the management and in various activities and projects of the UNESCO Chair [11] .

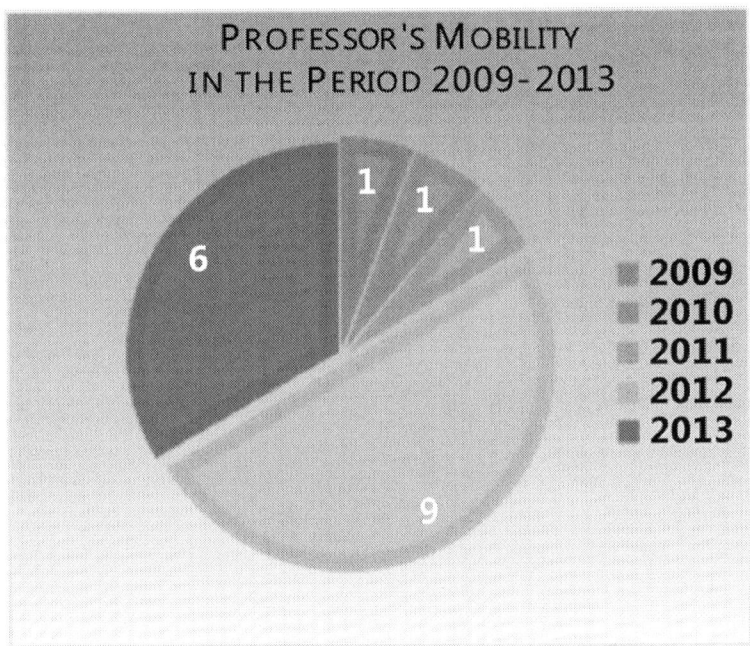

Figure 1. Professor's mobilityfor the period 2009-2013.

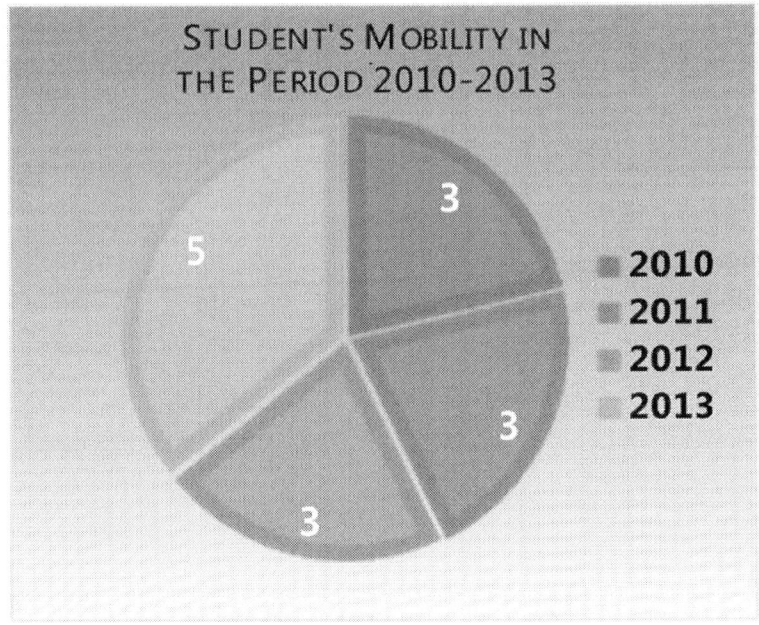

Figure 2. Student's mobility for the period 2010-2013.

We could summarize that for five years period from the establishment of Library Management Department in SULSIT we have achieved a presence in the single educational space in the field of Library and Information Sciences and we are highly motivated for future initiatives and partnerships.

4. CURRICULUM AND METHODOLOGY OF LIBRARY AND INFORMATION MANAGEMENT BACHELOR'S PROGRAM

The students which obtained the educational and qualification degree "Bachelor" in "Library and Information Management" in SULSIT must have a thorough scientific, theoretical and practical training in the specialty, which includes:

- Fundamental training on the nature and the diversity of libraries and other cultural institutions. History of the Field;

- Knowledge and understanding of the library cataloguing, acquisition and bibliographic processes and operations;

- Fundamentals of Library Management, Knowledge Management, Standardization and Quality Management in Library and Information Activities;

- Thorough knowledge of modern library and information technologies and their application to all facets of Library and Information Products and Services. Digitalization and Preservation of Library Funds;

- Theoretical and practical training in Information Resource Management, Information Retrieval, Assessing of Information Needs, Intellectual Property and Information Literacy;

- Specialized training in Library, Information and Project Management to acquire knowledge and skills in planning, organizing and controlling the overall operations of libraries, information centers, community centers and other cultural and public institutions;

- Basic knowledge of Marketing, Public Relations and Library Psychology for optimal organization of information and the use of methods to interact with audiences for library collections, services and events.

Experts should be able:

- To raise and resolve the duties arising in connection with the organization and ensure optimal use of information resources in libraries and other organizations with arrays;

- To develop and manage projects;

- To develop and implement programs for digitization and creating their own electronic resources;

- To organize and manage various initiatives to interact the various audiences;

- Carry out self-sociological, psychological and other research needed to analyze the quality of the services and the extent of their approval by the users, and to be able to improve the characteristics of information and communication forms used in the library, and to bring new services or forms;

- To collaborate effectively with national and international LIS community [12] .

Forms of education in educational and qualification degree "Bachelor" are regular and part time and are implemented in eight semesters. Table 1 presents the disciplines currently included in the curriculum for compulsory training of specialty "Library and Information Management" [13] .

Table 1. Compulsory subjects taught in the educational and qualification degree "Bachelor" in specialty "Library and Information Management" in SULSIT.

Discipline	Semester
History of the Cultural Institutions and Librarianship	
History and Theory of Culture	First semester
Information Resources	
Annotating and Referencing of Documents	
Information Systems	
Book Science	
Rare and Valuable Library Collections	
Bulgarian Literature	Second semester
English	
Second Foreign Language	
Sport	
National Bibliography	
Management and Library Collections Development	
Library Management	Third semester
Automated Libraries—Part 1. Cataloguing, Classification and Subject Indexing of Information Recourses.	
Automated Libraries—Part 2. Automation Library and Information Systems. OPAC, Union Catalogues. National and International Cooperation. Application of ICT in Libraries.	
Intellectual Property	Fourth semester
Information Services in Libraries	
Innovations for Preservation of Library Funds	
Library Marketing	
Information Literacy—Programs and Models	
Architecture of the Libraries	Fifth semester
Library Psychology	
Practicum in various types of libraries	
Information Management	
Knowledge Management	
Bibliometry	Sixth semester
Standardization in Library Activities	
Public Relations	
Library Legislation	
Quality Management of Library and Information Activities	
Digitization and Copyright	Seventh semester
Academic Writing	
System Analysis	
Project Management	Eight semester
Practicum and Fieldwork. Development of Bachelor Thesis.	

*Teaching English and sport is foreseen as a compulsory subject for each semester.

There is an option for the students to choose a discipline according to their interests (elective courses). Among the most popular are: Applied Software, Web Design, Content Management Systems, Intellectual

Property in Internet, Libraries and Local Authorities and Communities, European Communication Policy. Common culture among students is supplemented by the chosen of them optional subjects such as: Bibliotherapy, Access to information for people with special needs etc. Methodology of teaching the students is done except to traditional methods such as lectures and seminars, and through interactive and situational methods, application of the approach "learning by doing", "edutainment" and etc., which activates the students' participation. Individual forms of work such as term papers, presentations, individual assignments, communication in e-learning platforms—support the learning process and the current verification of students' knowledge. Library Management Department features with the active involvement of students in research projects and joint research papers for participation at scientific conferences and symposiums.

There are two initiatives of the young scientists in the Library Management Department, which are directed to the students. One of them is the work of students in relation to the Week Dedicated to The International Day of the Book and Copyright Day (23 April), with coordinator Dr. Lubomira Parizhkova, where students from all specialties take part into this initiative and have the opportunity to publish their first scientific research in the thematic research collection [14] . The second initiative is led by Dr. Tereza Trencheva and is dedicated to The International Intellectual Property Day (26 April), where the students have the opportunity to meet specialists from practice in the field of Intellectual Property [15] .

For self experience activities the students of Library and Information Management visited: libraries, museums, galleries, workshops for making paper and books, book fairs (with conducting inquiries), book premieres, meetings with authors, public lectures etc. Also, they participate in the initiatives of promotion of reading—marathon of reading, book crossing movement and in different forms of stimulation of children's reading.

Important project, which is creating the opportunity to use the acquired theoretical knowledge in a real working environment is the National Project' Students' Practices' of the Bulgarian Ministry of

Education and Science, implemented under the Operational Programme "Human Resources Development" of the EU. For the period July 2013-July 2014, more than 40 students of Library and Information Management Specialty conducted 240 hours internship in libraries, archives and information centers. Our students are actively involved in initiatives of the Bulgarian Library and Information Association and the Association of University Libraries, also. Our goal is nurture an active attitude towards the library profession and knowledge of the real challenges, problems and achievements in their future professional environment.

5. LIBRARY AND INFORMATION MANAGEMENT SPECIALTY WORLDWIDE

For the purposes of this research we have studied the higher education institutions in the world, which provide training in "Library and Information Management" (LIM). In this study, we aim to gain information in which educational and qualification degree is offered training in the specialty LIM and what are the main components in the content of the curriculum. As a source of information have been used official websites of the respective universities. The results are presented in a synthesized form in Table2

Table 2 shows that the universities conduct training in specialty 'Library and Information Management' mainly in master educational and qualification degree. The exceptions are Hochschule der Medien in Stuttgart, Germany; Peking University and The University of Hong Kong, China; Moscow State University of Culture and Arts (Department Information and Library Management), Russia; University Technology Mara (Faculty of Information Management), Malaysia.

Practical realization of bachelors graduated in 'Library and Information Management' in Bulgaria, is particularly suitable for libraries in small towns, whose staff is limited (and often reduced to a single librarian) and it is expected to be informed and competent in a wide range of

Table 2 . Worldwide Universities engaged in training in the specialty "Library and Information Management".

Country	Name of the University	Educational and Qualification Degree	Core Curriculum Elements
Australia	University of South Australia	Master (MA)	– Information Management – Accessing and Organizing Resources – Information Technologies
Canada	The University of British Columbia (School of Library, Archival and Information Studies)	Master	– Organization and Knowledge Management – Information Systems – Digital Libraries
China	Peking University	Bachelor (BA) Master	– Information Storage and Retrieval – Computer Networks – Information Economics – Management Information System – Library Management – Information Policy and Law
China	The University of Hong Kong	Bachelor Master	– Information Science – Science in the Field of Management
Denmark	Royal School of Library and Information Science	Master	– Theory of Production of Information and Knowledge – Analysis and Management of Information Resources – Information Policy and Strategy of the Organization – Planning, Management and Evaluation of Information Systems
Germany	Hochschule der Medien in Stuttgart	Bachelor Master	– Management in Human Resources, Organization, Marketing – Information and Media Market – Library Management System Database
India	Gujarat University	Master	– Management of Library and Information Centers – Leadership and Change Management – Information and Communications Technologies and Research
Malaysia	University Technology Mara	Bachelor	– Information and Library Management – Information, Information Technology and Libraries
Russia	Moscow State University of Culture and Arts	Bachelor Master	– Library and Information Management – Library Statistics, Legislation – Economics and Marketing of Library and Information Services – Information Technology Management – PR and Advertising in the Library
United Kingdom	University of Rhode Island (School of Library and Information Studies)	Master	– Management of Library and Information Services – Information Science and Technology – Types of Libraries and Library Processes and Services
	Robert Gordon University	Master	– Information Studies – Managing Information Services – Database Construction and Use – Knowledge Organization
USA	University of California Berkley	Master	– Information and Communications Technologies – Leadership – Computer Sciences – Psychology and Sociology, Economics, Business, Law, – Library/Information Studies

issues. As answer of these needs of library and information sphere, SULSIT offers training in Bachelor level—specialty "Library and Information Management" and in Master level—"Library, information and Cultural Management".

Diplomas and certificates that graduates of the corresponding universities receive are different, account the type of the training already acquired specialty and professional experience and length of the training course.

They can be:

- Bachelour diploma (training lasts 3 - 4 years);

- Graduate sertificate (training is done after obtaining a bachelor's degree or professional experience and usually lasts 6 - 8 months);

- **Postgraduate diploma** (training is done after obtaining a bachelor's degree or professional experience and usually lasts for 12 months);

- **Master** (training is done after obtaining a bachelor's degree and usually lasts 16 - 24 months);

- **Professional master** (training is done after obtaining a bachelor's degree and professional experience and usually lasts 12 - 18 months);

- **Graduate entry master** (training is done after obtaining a bachelor's degree in Library and Information Management or related specialty and usually lasts about 24 months);

- **PhD and other Doctoral Research Degrees** (training is done after obtaining a master's degree or completion of a bachelor's degree with special honors and usually lasts 2 - 4 years).

On the development of educational content of specialty "Library and Information Management" in SULSIT important influence had our joint work in the frame of Erasmus Intensive Programmes with our partners—the leading European universities, engaged in similar training programs in the field of LIS education. Table 3 presented the basic information about our partner's programs.

Table 3 . Partner universities of the Library Management Department in SULSIT, who carry out training in Library Management and/or Information Management.

Country	Higher School	Specialty and Level	Core Curriculum Elements
Austria	University of Vienna	Library and Information Studies MA	– Management (in Library, Information, Documentation) – System for Information and Documentation
Croatia	University of Zagreb	Library Science BSc, MA, PhD	– Bibliography – Computer Science – Library Management – Information Institutions Management Fundamentals – Libraries and Society
Czech Republic	Masaryk University Brno	Information and Library Studies BA, MA, PhD	– Library Processing and Services – Intellectual Property and Information Activity – Information Science – (Information Security, Information Systems, Information Education)
France	Institute of Technology Paris Descartes	DUT Universal Technical Diploma	– DUT Information Communication – Book Trade
Germany	Humboldt University in Berlin	Library and Information Sciences BA, MA Information Management BA, MA	– Digital libraries – Bibliometry, Infometrya, Scientometrics, – Information Policy, Ethics, Law – Information Retrieval and Exchange of Information – Communication and Management of Information and Knowledge
Germany	Hochschule der Medien in Stuttgart	Library and Information Management BA, MA	– Management in Human Resources, Organization, Marketing – Information and Media Market – Library Management System Database
Hungary	University of Szeged	Library and Information Science BA, MA	– Library Organization, Library System – Management of Technological Resources – Document Management – Legal Institutions of Documentation.
Lithuania	Vilnius University, Institute of Library and Information studies	Information Management in Libraries BA, MA, PhD	– Information and Library Service – Information Science – Culture Information and Communication – Creative Communication – Business Information Management – Public Relations
Poland	Jagiellonian University in Krakow.	Management and Social Communication BA, BSc, MA, PhD	– Information and Library Sciences – Culture and Media Management – Electronic Information Processing – Management – Social Policy
Turkey	Hacettepe University in Ankara	Information Management BA, MA, PhD	– Information Management – Information Architecture – Records and Database Management – Information Literacy, Standards and Cooperation in Information Management.

Review of the main directions in which are trained future library and information managers can generally be limited to the following: Information and Communication Technology and Resources; Information and Library Management (in particular—Management of Information Services, Management Information Systems, Personnel Management); Knowledge Organization. Academic subjects included in these directions are different for different universities.

6. FINDINGS FOR FUTURE CURRICULUM REVIEW

A detailed analysis of the information from this research leads to interesting conclusions that will be very useful for future curriculum review of the "Library and Information Management". We achieved some findings and solutions that we suggest to the attention of the LIS academic community worldwide as useful and relevant to the context of modern educational paradigm.

Special attention in this article we would like to devote to the fact that some academic subjects which are included in the curriculum of SULSIT and met student's interest do not exist or are very limited in the curricula of universities abroad. The development and integration of these disciplines in our educational plan is provoked by the interaction with prospective employers that define this knowledge as important and necessary for library and information professionals.

We draw attention to the following subjects:

- Intellectual Property;
- Standardization in Library Activities;
- Quality Management in Library and Information Activities;
- Library Psychology;
- Bibliotherapy.

6.1. The Training on Intellectual Property

In an knowledge based economy, an important place take experts who can interpret issues related to intellectual property, such as librarians and information specialists. It is they who are responsible for creating a policy of promoting understanding and resolving legal disputes and conflicts that are unique to this aspect of the Information Society [16] . One way to achieve this is through the educational impact of the curriculum on intellectual property. Intellectual property can be considered as an element of information literacy in university information environment, so to develop successfully students at the university, and in life, they must learn to use efficiently and effectively the wide variety of information and communication technologies for searching, finding, organizing, analyzing and evaluating the information they need. In addition, they need to understand the ethical use of information, including the violation of individual rights to intellectual property as plagiarism, use without permission of the author of works of literature, art, science, and also of patented inventions, industrial designs, indications (trademarks, geographical indications, domain names, companies). Finally, they should be able to systematize all this knowledge together to create an effective final product. This requires them to assemble the entire package of basic skills for research, technological skills, critical thinking and evaluation.

6.2. The Training on Standardization in the Library Activities

Knowledge of standardization and specific standards in all areas of human activity are not only useful but also necessary in the direct work of specialists and leaders in different organizational structures, including libraries. Learning of this discipline implies a successful career of students and saves them a lot of effort, time and resources to achieve and implement the requirements, rules, standards, approaches, methods and etc., established by the most highly qualified professionals at international, European and national level and involved in the development of these standards.

Among the most common library standards that future library managers in SULSIT learn about can be mentioned:

- BGS ISO 11690:2013 Information and documentation—Library performance indicators.

- BGS ISO 2789:2012 Information and documentation—International library statistics.

- BGS ISO 15511:2012 Information and documentation—International standard identifier for libraries and related organizations (ISIL).

- BGS ISO 5127:2009 Information and documentation—Vocabulary.

- BGS ISO 2709:2011 Information and documentation—Format for information exchange.

- BGS ISO 25577:2012 Information and documentation—MarcXchange.

- BGS ISO 3166-1:2006 Codes for the representation of names of countries and their subdivisions—Part 1: Country codes.

- BGS ISO 11799:2008 Information and documentation—Document storage requirements for archive and library materials.

Standardization in Library Activities discipline, taught in the sixth semester is fundamental for studying in the seventh semester Quality Management in Library and Information Activities.

6.3. The Training on Quality Management in Library and Information Activities

The educational content in this course is structured in two parts:

Fundamental part, which includes the following topics: the importance and relevance of quality management; historical development, total quality management, principles of quality management, principles for analysis, security, control, organization, systems for quality management standards, quality management.

Today more libraries develop and implement systems for quality management. The most commonly used worldwide are those based on a series of international standards ISO 9000 [17] . Therefore lead in the course lectures Quality management in library and information work is devoted to this particular system.Are also presented two other systems for quality management application in libraries—Balanced Scorecard (BSPE) and selforganizations under the criteria of the European Foundation for Quality Management (EFQM).

Specialized part, consistent with the professional field in which to be realized students in "Library and Information Management." This part of the curriculum content provides knowledge, documentation and implementation of systems for quality management in libraries and information centres, in view of their professional activity. The seminars helped to give young professionals skills to formulate the mission of the library, and the ensuing quality policy and objectives to achieve quality. Special emphasis on the brainstorming technique of K. Ishikawaas an opportunity to identify the causes of a problem and their elimination [18] . Students are introduced to the documentation requirements of the quality management system, and sample content of the basic documents of the Quality Management System (QMS), which adapts to the specific type of library.

6.4. The Training on the Library Psychology

The problem of improving the quality of library services at the level of librarians' professionalism today is extremely topical. These factors increase their need for psychological and pedagogical knowledge.

The course on Library Psychology aims to equip students with true understanding and knowledge of the nature, functions and principles of Psychology, as part of the activities in the library environment to provoke new ways of working and communicating with users of library services. The course content is divided into two main modules. In the first module "Library Psychology" is considered library psychology as a scientific and academic discipline, traced the historical development of library psychology, the role of books in modern society. In the second module "Psychology of Communication in the Library" focuses on issues of personal and professional qualities of the library specialist; the psychology of communication between Librarian and Reader; the peculiarities of psychological climate in the library team and theformation steps of an effective team in the library; the nature and style of the library manager. Library Psychology has significant potential, not only theoretically, but practically oriented knowledge useful in the work of librarians from different types of libraries.

6.5. The Training on the Bibliotherapy

The course provides theoretical and practical knowledge to the students about the nature and application of Bibliotherapy in libraries. Students learn the historical and scientific development of Bibliotherapy. The Bibliotherapy brings together at least three fields of knowledge—a Science Studying the Book (Book Science, Literature, Library), the Science of the Human Soul, Targeted Bibliotherapy (Medicine, Psychiatry, Psychotherapy, Rehabilitation) and the Science of reading, (Ensuring Efficiency of Bibliotherapy, Psychology of Reading, Reading Instruction), which makes it extremely useful in the training of students. The lectures aim to examine possibilities which reveal Bibliotherapy by presenting its various forms, methods and functions. The students learn about new and alternative forms of employment applicable in the library environment.

The course includes the following three aspects: first module "Origin and Nature of Bibliotherapy" which focuses on the scientific development of bibliotherapy, definition of bibliotherapy, determining bibliotherapy, the types and purposes, her functions and tasks. In the second module, "Applications of Bibliotherapy" are tracked innovative techniques used by library professionals, bibliotherapy as a form of working with children with special needs. The third module, "Fairy Tales Therapy" is a brand new in Bulgaria and for the first time available as part of an academic course. In it, students get acquainted with the possibilities of therapy tales methodological model of working with stories, kind words and ways of working with them.

Bibliotherapy is dynamic and flexible science for support, encouragement, challenge, identifying and addressing a number of significant problems in modern society. It is a mechanism for painless reach the soul and the possibility of achieving lasting results and as such may be extremely useful in the training of future specialists.

7. DISCUSSION

In the present article, we pay attention to the above mentioned disciplines, as the analysis of the curricula of many universities around the world has shown that these disciplines are still rarely studied. In Table 2 and Table 3, Universities which studied one or more than one of the subjects on which we focus are present. Relatively new disciplines are presented by us, consistent with the requirements of the modern information society, and the role and functions of libraries in it. We believe that along with the subjects proved important for the preparation of future LIS professionals as library management, library legislation, library marketing, etc. would be appropriate curricula to attend and complementary knowledge of library professionals, consistent with the current state of society and libraries. The information presented can be used by other universities in Europe and beyond for comparison and analysis.

8. CONCLUSION

Through active partnerships with foreign universities, innovative teaching methods appropriate to the training of future specialists disciplines, and the continuous enrichment capacity and experience of the faculty, trained students in specialty "Library and Information Management"—we aim to be one of the leading educational institutions and adequate to the modern needs. We hope that shared insights and achievements will provoke an open professional dialogue for the enrichment of library and information academic programs. This collaboration could be going on together with the LIS community dialogue about IFLA Trends and the future of libraries and librarian profession in general [19] .

Acknowledgements This publication has been realized under the project "Development of an information environment to motivate and stimulate young researchers in SULSIT" by Contract: BG051PO001-3.3.06-0055. The project is funded by the Operational Programme "Human Resources Development" SchemeGrantsBG051PO001-3.3.06

"Support for the development of PhD students, post graduate students and young scientists" funded by the European Social Fund of the European Union.

REFERENCES

1. Kajberg, L. (2008) The European LIS Curriculum Project: Findings and Further Perspectives. Zeitschrift für Bibliothekswesen und Bibliographie, 55, 184-189.

2. Juznic, P. and Badovinac, B. (2005) Toward Library and Information Science Education in the European Union: A Comparative Analysis of Library and Information Science Programmes of Study for New Members and Other Applicant Countries to the European Union. New Library World, 106, 173-186.

3. Georgy, U. (2009) Internationalization of LIS Education within the Bologna Process—Mobility and Flexibility. IFLA, Milano. http://www.uas7.org

4. Kawalec, A. (2014) Education, Competencies, Skills in the Field of Information and Library Science in Europe. Library (R)evolution: Promoting Sustainable Information Practices: Abstracts of the 22nd International BOBCATSSS symposium, 29-31 January 2014, Barcelona, 16. http://bobcatsss2014.hb.se/wp-content/uploads/2014/01/Bobcatsss-abstract-book.pdf

5. LIS Education in Europe. http://www.iva.dk/LIS-EU/project.asp/

6. Kajberg, L. and Lorring, L. (2005) European Curriculum Reflections on Library and Information Science Education. The Royal School of Library and Information Science, Copenhagen, 241 p. http://www.library.utt.ro/LIS_Bologna.pdf.

7. Pannier, G., Wilhelm, H. and Todorova, T. (2011) IPBib: Das Grimm-Zentrum-(k)ein Bibliotheksmarchen. Mobility and Innovation in the European Context. Proceedings of Evaluation

Conference on ERASMUS Intensive Programmes, Federal Ministry of Education and Research, Bonn, 2011, 19-21.

8. Project Website (2013) Erasmus Intensive Programme "Library, Information and Cultural Heritage Management— Academic Summer School". http://libcmass.unibit.bg/

9. Todorova, T. (2012) Library, Information and Cultural Heritage Management: Textbook. Za bukvite-O Pismeneh, Sofia, 246.

10. Todorova, T., et al. (2013) The Changing Role of the Manager in the Digital Era: Findings from Erasmus IP LibCMASS 2012 Project in from Collections to Connections: Turning Libraries "Inside-Out". Proceedings of the 21th International BOBCATSSS Conference on Information Science in Ankara, Ankara, 23-25 January 2013, 202-204. http://bobcatsss2013.bobcatsss.net/proceedings.pdf

11. (2013) UNESCO Chair ICT in Library Studies, Education and Cultural Heritageat State University of Library Studies and Information Technologies. http://unesco.unibit.bg/

12. Yankova, I. (2009) Library Management and New Challenges in Digital Era in Libraries and Their Clients: Free or fee Services Supporting Social Communication in Digital Era. Proceedings of the ePublications of the 15th Jubilee International Conference of JU LIS Institute in Kraków, Poland, 76-80. http://skryba.inib.uj.edu.pl/wydawnictwa/e06/yankova.pdf

13. State University of Library Studies and Information Technologies. http://www.unibit.bg/learning-activity/bachelor/bachelor-plans.

14. Parizhkova, L. (2013) The Book Our More Sensual Present: Textbook. Za bukvite-O Pismeneh, Sofia, 248.

15. Trencheva, T. and Eftimova S. (2013) Intellectual Property at the Universities—Creativity: The Next Generation. Textbook. Zabukvite-O Pismeneh, Sofia, 206.

16. Joint, N. (2006) Teaching Intellectual Property Rights as Part of the Information Syllabus. Library Review, 55, 330-336.

17. Balague, N. and Saarti, J. (2011) Managing Your Library and Its Quality: The ISO 9001 Way. Chandos, Cambridge.

18. Ishikawa, K. (1990) Introduction to Quality Control, 3A Corporation, Tokyo.

19. (2013) IFLA Trend Report. http://trends.ifla.org/

Chapter 15

Mapping of an Aptitude of Library Science students in Relation to Some Variables

Dr. Nileshkumar M. Kantaria

Smt. M. M. Shah College of Education, (Gujarat) India.

ABSTRACT

This study was undertaken to measure/mapping an Aptitude of library science students of Library Science Department Gujarat State. A total of 184 responses were collected in this study. In this study, researcher intended to find out the effect of variables i.e. Gender, Social Group and Faculty on Aptitude Test Score (ATS) of library science students. Librarian Aptitude test is constructed and standardized to gauge one's overall academic and professional competency. The finding of this study is showing that only one variable, Faculty of library science students' is effect on their Aptitude. LIS departments of all university should re think their policy for admission in this profession and re design the curricula with current scenario.

Keywords: Aptitude, Aptitude Test, Aptitude Test Score (ATS), Librarian Aptitude, Library Science.

INTRODUCTION:

It is a well-established fact that to succeed in any profession of today's highly competitive era, expertise knowledge of that particular profession is indispensable. For achieving progressive success in any profession, some specific interests, abilities and aptitudes well related with that particular profession is necessary. If we look at the history of human development, man has progressed through different types of societies. First, it was an agriculture based society, next came the era of industrial revolution in which the society became an industry-based one and now, at present, the third era has begun in which knowledge and information are most important for any type of development. The impact of this knowledge-information based era can be seen in the process of curriculum development for the degree of Library and Information Science. The curriculum has been enriched and expanded regarding the duties and responsibilities of the Librarian with a view to cope with the demands of the changing time. In this regard the person who has no interest in Library science education, has no interest in this profession, has no interest in knowing new techniques of information science and whose attitude is not positive towards users of library, if such person enters the profession of Librarianship then the next generation will suffer from his or her services. If we can confidently say the person who joins the profession of Librarianship has essential qualification and aptitude for this profession, then the output of Library science profession will be bright. However, today the admission of students in Library science profession is only on the basis of merit of his graduation marks. Many questions can be raised up regarding the reliability of today's examinations and evaluation systems. Therefore, to say that students who have high merit will be successful as a Librarian is negligence of most important ability to succeed in this profession. Therefore, the researcher has selected this problem. In this study, the researcher is going to construct and standardize an aptitude test for the Librarianship profession and study the effect of some variables on Librarian aptitude test scores.

In present, study the term Aptitude is used in its wide and familiar meaning. Aptitude means the collection of qualities, characteristics and abilities which are God gifted or acquired and which show that if a person gets appropriate training in the relevant field he will get excellence in that particular field.

Statement of Problem:

The Researcher has formulated the problem for the present study as below **"Mapping of an Aptitude of Library Science students in Relation to Some Variables"**.

Review of Literature:

According to L. P. Mehrotra "One of the essential aspects of a research is to review the related literature. The investigator should know that his problem is not absolutely new but a lot of work has already been done on the problem which he proposes to study and therefore his effort should be acquaint himself with all the connected literature contributed previously by other investigators as far as possible. Such a systematic, through and relevant review of material promotes a greater understanding of the problem and resumes the avoidance of unnecessary duplication. It also helps him to make a comparative study of his findings with those of others and thus evaluate and interpret their significance."

In this study, the Researcher had tried to find out some new facts about the problem, with review of some research work all ready done. I have read and referred some related researches has been conducted in the past, are as under.

Construction and Standardization of Secondary Teacher Aptitude Test, Ph.D. Thesis by J. J. Dixit, Education Dept., Bhavnagar University, Bhavnagar. 2007.

Construction and Standardization of an Aptitude Test for Primary School Teachers, Ph.D. Thesis by M. U. Tamaliya, Education Dept., Saurashtra University, Rajkot. 2001.

Construction and Standardization of a Scientific Aptitude Test in Oriya for the 10th Class Students of Orissa, Ph.D. Thesis by Banmalidas, Education Dept., Kurukshetra. University, 1987.

The Construction and Standardization of a Musical Aptitude Test for Gujarati Children, Ph.D. Thesis by D. S. Shukla, Gujarat. University. 1987.

Construction and Standardization of Mechanical Aptitude Test in Oriya for 10th Class Students of Orissa, Ph.D. Thesis by S. K. Swain, Education Dept., Kurukshetra. University, 1986.

Need for the Study:

Libraries are essential to both civilization and democracy: they gather the former's collective intelligence, and facilitate its access. It is an accepted fact that integral and positive development of any society is highly dependent on its libraries, ranging from, school libraries, college libraries, university libraries, public libraries and research and other special libraries. Libraries have a recognized social function in making knowledge publicly available to all. They serve as local centers of information and learning, and are local gateways to national and global knowledge.

However, today the admission of students in Library science profession is only on the merit of his graduation marks. Many questions can be raised up regarding the reliability of today's examinations and evaluation systems. Therefore, to say that students who have high merit will be successful as a Librarian is negligence of most important ability to succeed in this profession. Therefore, the researcher has selected this problem.

While no test is an infallible gauge of one's ability, standardized aptitude tests provide students, school admission boards, and potential employers a basic idea of where an individual's weaknesses and strengths lie, as well as where they rank among their peers. It is for this reason such tests are a beneficial tool in the business and academic worlds.

Present study will show the current situation of Aptitude Score of library science students of Gujarat State. Findings of this study will show which independent variables effects on Aptitude. We can give him guidance and counseling according their ability and interest.

Objectives:

To find out library science students' aptitude test score

To find out library science students' aptitude test score with the context of their gender

To find library science students' aptitude test score with the context of their social group

Study of effect of faculty's on aptitude of library science students'.

Variables:

In this study, following dependent and independent variables were taken for the study of effect on aptitude of library science students'.

Dependent Variables: Aptitude Test Score of library science students'.

Independent Variables:

NO.	VARIABLE	LEVEL OF VARIABLE
1	Gender	(1) Male (2) Female
2	Social Group	(1) Reserved (2) Non Reserved
3	Faculty	(1) Arts (2) Other

Hypotheses:

In this study, following null hypotheses were framed based on independent variables, which were taken for this study.

There will be no significant difference between the mean scores of Aptitude Test Score of library science students' with respect to their gender.

There will be no significant difference between the mean scores of Aptitude Test Score of library science students' with respect to their social group.

There will be no significant difference between the mean scores of Aptitude Test Score of library science students' with respect to their faculty.

Methodology:

The present study was carried out using survey as the research method. **Sample:** The present study was conducted on a sample of 184 library science students of Library Science Department of Gujarat State. **Research Tools:** In this study, Aptitude Test constructed and standardized by Researcher (Dr. Nileshkumar M. Kantaria) was used for data collection. **Data Collection:** The data was collected in the normal classroom situation (condition) by investigator himself from the Library Science Department of Gujarat State. Librarian Aptitude Test was taken of library science students. **Data Analysis:** The response of library science students on Aptitude Test was evaluated. Mean, Standard Deviation and t-test were applied for testing hypotheses. The computer programme MS-Excel was used to analyze the collected data.

Result and Discussion:

There is three hypotheses had been constructed to know any effect of independent variable i.e. Gender, Social Group and Faculty on Aptitude Test Score of library science students. For statistical calculation, all hypotheses were converted into null hypotheses.

Mean, Standard Deviation, t-Value of each respondent group had been found out, and hypotheses had been tested at 0.01 and 0.05 level of Significant.

Analysis and interpretation of Aptitude Test Score of library science students in the context of Gender is shown below.

Table – 1: Mean, SD and t – Value scores on Aptitude Test of Male and Female Library Science students

Variable	Respondent	Mean	SD	σ D	't' Ratio	Significance Level
Male	51	255.31	24.19	3.79	0.10	Not Significant
Female	133	254.93	19.71			

In above table mean of male library science students is 255.31 and mean of female library science students is 254.93. Standard error of Mean is 3.79. To find out of significance of means difference, t–Value is calculated. t-Value is 0.10 that is not significant at any level.

Result of tasted hypothesis show that groups of male and female library science students are equal in Aptitude Test Score. Thus in present study this hypothesis is not rejected.

Analysis and interpretation of Aptitude Test Score of library science students in the context of Social Group is shown below.

Table – 2: Mean, SD and t – Value scores on Aptitude Test of Male and Female Library Science Students

Variable	Respondent	Mean	SD	σ D	't' Ratio	Significance Level
Reserved	110	253.74	21.65	3.10	1.04	Not Significant
Non-Reserved	74	256.97	19.93			

In above table mean of reserved library science students is 253.74 and mean of non- reserved library science students is 256.97.

Standard error of Mean is 3.10. To find out of significance of means difference, t–Value is calculated. t-Value is 1.04 that is not significant at any level.

Result of tasted hypothesis show that groups of reserved and non-reserved library science students are equal in Aptitude Test Score. Thus in present study this hypothesis is not rejected.

Analysis and interpretation of Aptitude Test Score of library science students in the context of Faculty is shown below.

Table – 3: Mean, SD and t – Value scores on Aptitude Test of Male and Female Library Science Students

Variable	Respondent	Mean	SD	σ D	't' Ratio	Significance Level
Arts Faculty	132	252.92	19.98	3.59	2.09	0.05
Other Faculty	52	260.42	22.64			

In above table mean of arts faculty library science students is 252.92 and mean of other faculty library science students is 260.42. Standard error of Mean is 3.59. To find out of significance of means difference, t–Value is calculated. t-Value is 2.09 that is significant at 1.5 level.

Result of tasted hypothesis show that group of other faculty library science students superior to arts faculty library science students in Aptitude Test Score. Thus, in present study this hypothesis is rejected.

Findings:

The findings of the study are as under.

- There is no significant difference in means score of Aptitude Test between male and female library science students.

- There is no significant difference in means score of Aptitude Test in the context of Social Group of library science students.

- There is significant difference at 0.05 levels in means score of Aptitude Test between Faculty in favour of other faculty library science students.

CONCLUSION:

This study was about Mapping of Aptitude of Library Science students in Relation to Some Variables. Librarian Aptitude Test constructed and standardized by Researcher (Dr. Nileshkumar M. Kantaria) was used for data collection. The finding of this study is showing that one variable, Faculty of library scenic students is effecting on their Aptitude. On last, Aptitude Test is helpful to decide for admission in Library & Information Science Profession. It is also helpful for faculties to give guidance and counseling to the students.

SCOPE FOR FURTHER RESEARCH:

There is no research is last, but it left behind many questions for further study. That is why always remains scope for further research. The researcher suggests that the Relation between Aptitude Score and Job Satisfaction of Librarian should be investigated. Effect of other variables on Aptitude of library science students is also interesting for the further study. It will be interesting to comparative study between Aptitude and Attitude of Library Professionals.

REFERENCES:

1. Cronbach, Lee J. (1949), *Essentials of Psychological Testing.* Harper & Brothers, New York. Dawra, Manisha (Ed.). (2003), *Encyclopedia of Modern Library and Information Science. Library Science and Theories. Vol. 3.* Rajat Publication, New Delhi.

2. Dawra, Manisha (Ed.). (2003), *Encyclopedia of Modern Library and Information Science. Library Science and Theories. Vol. 4.* Rajat Publication, New Delhi.

3. Kumar, P. S. G. (Comp.). (1987), *Research in Library and Information Science in India*. Concept Publishing Company, New Delhi.

4. Mahapatra, P. K. (2002), *Human Resource Management in Libraries*. Ess Ess Publication, New Delhi.

5. Mouly, George J. (1964), *The Science of Educational Research*. Eurasia Publishing House (pvt.) Ltd., New Delhi.

6. Seetharama, S. and Karisiddappa, C. R.(Ed.). (1993), *Current Research in Library & Information Science*. R B S A Publishers, Jaipur.

7. Shukla, Bana Bihari. (1987), *Library and Community, Library Administration and Management*. Bharti Publication, Cuttack.

Index

(E-)learning, 48

A
Adaptable Cycle, 199, 200, 202
Aptitude, 251, 253, 254, 255,
  256, 257, 258, 259
Aptitude Test Score (ATS), 251

B
Bachelor's Degree, 228
Bonding, 200
Budgets, 200

C
Community, 19, 200, 204, 206,
  218, 224, 260

D
Data integrity, 74
data standardization, 73, 74,
  75, 76, 77, 78, 79, 80, 81
digital library, 63, 73, 74, 75,
  76, 77, 78, 80, 83, 129, 130,
  137, 166, 167, 168, 169, 170,
  171, 172, 176

E
ETDs, 74, 78, 79

F
facilities, 19, 29, 34, 38, 88, 89,
  92, 93, 96, 100, 101, 107,
  114, 120, 185
Frank Lloyd Wright, 29, 30, 34,
  35, 39, 40, 41

H
Higher Education, 80, 81, 82,
  149, 177, 228
History, 12, 24, 26, 40, 67, 234

I
information literacy, 1, 2, 3, 4,
  5, 7, 53
Information Management, 66,
  69, 70, 73, 87, 93, 97, 227,

L
Legislation, 12
libraries, 1, 2, 3, 4, 6, 7, 8, 9, 11,
  12, 13, 14, 15, 16, 17, 19, 20,
  238, 243, 244, 245, 246, 247,
  254
Library Management, 126, 197,
  234, 237, 238, 241, 249

Library Science, 11, 24, 67, 91, 248, 251, 253, 256, 257, 258, 259
literacy, 1, 2, 3, 4, 5, 6, 8, 9, 30, 55, 56, 62, 66, 69, 71, 72, 242

M
Mean response delay, 142
mould growth, 149, 150, 151, 152, 153, 154, 155, 156, 157, 161, 162
Mould Sampling Method, 152, 153

N
navigation support, 130, 140, 141, 146

P
Patrons, 200, 214, 215
portal, 69, 130, 132, 133, 134, 135, 136, 139, 140, 141, 142, 145, 166
Public Engagement, 200
Public Library, 9, 109, 126, 201, 203, 205, 206, 213

Q
qualitative assessment, 149, 150, 151, 154, 155, 162
Quality of Service, 165

R
Request delivery rate, 142
Request success rate, 142

resources, 8, 16, 17, 18, 19, 20, 221, 228, 235, 243
Return on Investment, 200

S
school research center, 88
SLAs, 165, 167, 168, 169, 170, 171, 172, 173, 174, 175, 176, 178
Social Media, 200
social navigation, 130, 131, 139, 140, 141, 143, 144
spatial and Landscape arrangement, 30
SULSIT, 227, 228, 230, 231, 234, 236, 239, 240, 241, 242, 243, 247

T
technological fluency, 1, 3, 4, 5, 6, 7
Tertiary Libraries, 181, 198
Third-party Sourced Services, 165
TUHH library, 57
Turkey, 73, 74, 78, 79, 80, 81, 83, 232

V
visual inspecti, 150

W
Waldens' Paths, 138